Food, Consumers,
and the
Food
Industry

Catastrophe or
Opportunity?

CRC Series in
CONTEMPORARY FOOD SCIENCE

Fergus M. Clydesdale, Series Editor
University of Massachusetts, Amherst

Published Titles:

America's Foods Health Messages and Claims:
Scientific, Regulatory, and Legal Issues
James E. Tillotson

New Food Product Development: From Concept to Marketplace
Gordon W. Fuller

Food Properties Handbook
Shafiur Rahman

Aseptic Processing and Packaging of Foods:
Food Industry Perspectives
Jarius David, V. R. Carlson, and Ralph Graves

The Food Chemistry Laboratory: A Manual for Experimental Foods,
Dietetics, and Food Scientists
Connie Weaver

Handbook of Food Spoilage Yeasts
Tibor Deak and Larry R. Beauchat

Food Emulsions: Principles, Practice, and Techniques
David Julian McClements

Getting the Most Out of Your Consultant: A Guide
to Selection Through Implementation
Gordon W. Fuller

Antioxidant Status, Diet, Nutrition, and Health
Andreas M. Papas

Food Shelf Life Stability
N.A. Michael Eskin and David S. Robinson

CRC Series in
CONTEMPORARY FOOD SCIENCE

Food, Consumers, and the Food Industry

Catastrophe or Opportunity?

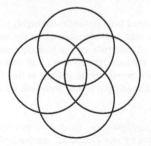

Gordon W. Fuller

CRC Press
Taylor & Francis Group

Library of Congress Card Number 00-046820

Library of Congress Cataloging-in-Publication Data

Fuller, Gordon W.
 Food, consumers, and the food industry : catastrophe or opportunity? / Gordon W. Fuller.
  p. cm. — (CRC series in contemporary food science)
 Includes bibliographical references and index.
 ISBN 0-8493-2326-6 (alk. paper)
 1. Food. 2. Consumer protection. 3. Food industry and trade. I. Title. II. Series.

TX353 .F835 2000
363.8—dc21   00-046820
                      CIP

CRC Press
Taylor & Francis Group
6000 Broken Sound Parkway NW, Suite 300
Boca Raton, FL 33487-2742

© 2001 by Taylor & Francis Group, LLC
CRC Press is an imprint of Taylor & Francis Group, an Informa business

First issued in paperback 2019

No claim to original U.S. Government works

ISBN-13: 978-0-367-45538-5 (pbk)
ISBN-13: 978-0-8493-2326-3 (hbk)

Preface

For nearly 50 years I have worked exclusively in the agricultural and food industries. My work has been in government as a research analyst; in industry as a hands-on technologist in product development, with a research association in charge of meat products research, and later as vice president, Technical Services, of a multinational food company; in academia as an associate professor teaching and directing research in added value poultry products; and in private practice as a food consultant. During the course of this work, I visited growers, food and ingredient manufacturers, and government research establishments in countries in North and South America, Asia, and Europe. I have lectured or conducted training courses in Canada, the United States of America, England, Ecuador, The Netherlands, China and Germany.

Such opportunities as these travels permitted allowed me to meet and discuss food issues with many distinguished men and women involved in various activities within the food microcosm. Knowingly or unknowingly they helped shape many of my thoughts and concerns that have been presented in this book. I thank all of them.

I have deliberately tried to steer a course between presenting a technical discourse on food issues in the third millennium and creating a more readable, non-technical presentation of the concerns that will confront future generations. Support material for the issues presented has come from the public media, popular scientific magazines and peer reviewed journals. My colleagues may find this approach too non-technical. However, it was necessary that all, scientists, lay business people and the general public see how we, the technologists and food business community, are perceived in the public media as well as in technical journals. For those in the non-technically trained general public who are concerned about their food I trust the book is not too technical and may stir some

available to all at a price they can afford but continue to be available to future generations.

I would, in particular, like to acknowledge the assistance of my son, Grahame Fuller, who prepared the figures.

The Author

Dr. Gordon W. Fuller has had extensive experience in industry, in academia and in government, all of which was in association with food and agriculture. In government, he worked as a research analyst on the characterization of citrus flavors using gas chromatography with the Food and Drug Directorate in Ottawa, Canada.

Industrial activities included work on chocolate and chocolate products with the Nestlé Co , Fulton, NY; tomato, fruit, and canned soup products with the Mellon Institute for Industrial Research, Pittsburgh, PA, for the H. J. Heinz Co.; and fresh and preserved meat products with the Food Research Association, Leatherhead, U.K. This product experience was broadened when as vice president of technical services, Imasco Foods Limited, he was responsible for corporate research and development over a product line that included canned goods, and bakery operations for tortillas and frozen foods.

In academia, Dr. Fuller was an associate professor in the department of poultry science at the University of Guelph. Here he taught and conducted research on poultry and egg products. He was also an external lecturer at both Concordia and McGill Universities in Montreal.

He is the author or co-author of several papers in scientific journals and food trade magazines. He is the author of *New Food Product Development: From Concept to Marketplace,* published by CRC Press (1994) and *Getting the Most Out of Your Consultant* also published by CRC Press (1999). He is a Fellow of the Institute of Food Science and Technology (U.K.), a charter member of the Institute for Thermal Processing Specialists, and is a member of the Institute of Food Technologists and the Canadian Institute of Food Science and Technology.

Table of Contents

Dedication

To my wife, Joan, for her proofreading and suggestions for improvements and clarification of my writing, but most of all for her patience, understanding, and encouragement.

Chapter 1

The Present Status

"Organic life, we are told, has developed gradually from the protozoon to the philosopher, and this development, we are assured is indubitably an advance. Unfortunately, it is the philosopher, not the protozoon, who gives us this assurance."

Bertrand Russell, *Mysticism and Logic* (1917)

Introduction

The end of the second millennium and the start of the third have prompted a number of speculations about the "greatests" of the past century or even of the past millennium. What were the greatest scientific discoveries? The greatest historical events changing civilization? Who or what were the greatest newsmakers, politicians, hockey players, fastest runners, best books, etc. ad nauseum. Russell, an eminent philosopher, as noted above pokes a bit of fun at progress such as philosophers are apt to judge it by suggesting progress be viewed from a different point of view.

In the food microcosm, does it mean that progress is equated with foods that are more natural; more flavorful; healthier; more 'ready-to-serve'; or genetically modified for more flavor, for greater pest or pesticide resistance, for greater hardiness; or for more cancer-reducing factors; or for fewer allergic factors? Technocrats may certainly think so but it seems

Genetic engineering technologists endeavoring to produce peanut varieties that can grow in southern Ontario above the 42nd parallel might be considered by many to be attempting what I have elsewhere termed "little Jack Horner research." Little Jack Horner in the nursery rhyme stuck his thumb into his Christmas pudding and pulled out a plum and said 'what a good boy am I'. Is there a need for such esoteric research? Is there an economic demand that motivates such an endeavor? Or is this merely a manifestation of need on the part of scientists to demonstrate how clever they are? Might scientists next attempt hybrid varieties of crops such that future generations might expect coconut plantations in the frozen tundra or coffee plantations on the prairies to replace wheat? Or, as one letter-to-the-editor writer queried, "might we not expect edible poison ivy in the same manner as inedible rapeseed oil was made into edible canola oil?"

This book is concerned with those issues facing the food microcosm from the primary producer or gatherer to the consumer. Today, at the end of the second millennium and the start of the third, is as likely a place as any to review developments in the food microcosm and to analyze issues that have arisen for their import in the young millennium. These issues must be resolved in some acceptable, not necessarily logical, fashion. At the very least, they must be addressed in a way acceptable to all the consuming public, who are not always logical, and they must be acceptable to several other constituencies such as politicians, farmers, environmentalists, and legislators with vested interests.

A Prelude

As many Christian philosophers claim, the first Christian millennium ended with two prevailing schools of thought. Both schools believed that the year 1000 A.D. would bring some profound change: The passing of the old millennium was the end of the world as it was known. The doom-and-gloom group believed fervently that Judgement Day would come at midnight, 1000 A.D.; the world would come to an abrupt end and all would be judged. The other group, the happy holiday campers, believed, equally fervently, that Jesus Christ would return to earth for a 1000-year period of world peace, prosperity, and happiness.

Both camps were to be disappointed, and today a certain amount of chuckling may be permitted at such naïve thinking.

Nevertheless, the approach of the third millennium of the Christian era

hand, there were those experts, equally gifted, who maintained that there would be no problems at all or only very minor glitches. Again, a dichotomy of opinion, as appeared a thousand years before, was prevalent.

There were stories in newspapers of concerned citizens stockpiling food and water: Two years (!) supply of food and water were recommended. Some alarmists established communes in remote locales where they were going to build fortified communities and, incidentally, were selling land in these remote communities. Here, in these sites far from urban centers, entrepreneurs set up colonies of like-minded individuals who were prepared to brave out the new millennium hunkered down in their fortress communities prepared for a back-to-the-earth way of life. Sales of dehydrated foods soared. So the availability of food and water were concerns of many as the second millennium drew to a close.

Thus began the third millennium.

The Pressure and Power of Food in Society

Thoughts of food hardly entered the heads of either those who expected an end to the world or those who believed in a benevolent existence to come in the passing over into the second millennium. After all, there would be no need for sustenance in either a thousand years of Heaven on earth where a benevolent Creator would provide for all their needs or a Hell somewhere else.

Food did concern people at the passing over into the third millennium: Stockpiling resulted from fear of the chaos promised by some. Food will have a major role to play in the coming millennium as it has in past history but perhaps for different reasons than it has in the past. Wars have been fought over productive farming lands as well as over control of the oceans that can yield food. Wars have been won and lost over the availability of food. Food, its lack or even its quality, can be a determinant in the physical, social, and moral health of societies. Riots have been caused by poor quality or lack of variety of food in prisons and in the military. Food, the 'bread' along with the circuses that one reads about in history books, kept the Roman populace under a degree of control in ancient Rome. A rising cost of food can spark unrest among the general population as it did, for example, when the Mexican government removed its subsidy on tortillas on January 1, 1999. This caused prices of this basic of the Mexican diet, especially the diet of the less privileged Mexicans, to rise. The government

4 ■ Food, Consumers, and the Food Industry

ecosystems by overzealous agricultural practices. Over-fishing has seriously depleted commercial fish stocks, harmed the economies of maritime nations, caused bitter rivalries between countries and even led to armed interventions.

Food trade disputes have erupted in Europe over importation of grapes for winemaking between France and Italy; in North America over wheat shipments between Canada and the United States of America, in Japan over rice imports between the government and Japanese rice growers, between Canada and Spain and between Iceland and the UK over fishing rights. Various news media (December 1998) described an impending trade war between the U.S. and the European Community over bananas — Europe was buying its bananas exclusively from its former colonies while the U.S. was promoting bananas from the Latin American countries where the growing of bananas is largely under the control of large American companies. It is still largely unresolved as this is being written in the first year of the new millennium.

Meanwhile, famine is rampant in many parts of the world. Weather, either too wet or too dry, has had adverse affects on many growing areas usually considered to be 'breadbaskets' to feed peoples in less productive regions. The world's population shows no signs of stopping its meteoric growth rate. Despite famine, war and disease, the world's population adds more and more mouths to be fed from a limited resource base: the earth, with its oceans, lakes and rivers harmed by the pollution of industrial growth and arable lands that are being encroached on by the growth of cities.

The great age of the voyages of exploration brought about a greater variety of foods to the European diet. Appert's success with preservation of food by thermal processing removed or greatly reduced the dependency on the seasonal availability of food.

The latter half of the 1900s ushered in the age of space exploration. It became an age of excitement. But food, common ordinary food, may become a determinant in space exploration, limiting distances and duration of space flights: Astronauts, as Napoleon noted for his armies, cannot go beyond their available food supply. In space, there are no conquered lands from which food resources can be commandeered or requisitioned by exploring astronauts.

If, as it has been forecast, only 15% of an astronaut's calories can come from Earth, astronauts will have to devote a large part of their precious cargo space for the paraphernalia necessary to grow, process, and store food safely. This is a challenging endeavor.

■ Space and weight allotments, already very restricted in spacecraft, must be given over to this person's personal living quarters and to activities for the production of food in space. A brief catalogue of needs includes seeds, seedling beds, a suitable light source, proper temperature control, a source of plant nutrients, microorganisms for food or to convert plant waste to compost (and water) and precious water. And still this raw plant material must be converted to appetizing food. If small animal husbandry or insect culture for food is included, a large space and weight allowance must be given for the paraphernalia for food production in space.

For the success of any long-term space exploration, such agro-astronauts will be a necessary and integral part of any expedition.

Agro-astronauts will need to be trained in such skills as:

■ The growth, maintenance, and propagation of food plants and perhaps even for care and rearing of small animals, insects, and microorganisms for food use;

■ The preparation of tasty, nutritious, and safe (of utmost importance on long space flights) foods from these resources; and

■ The recycling of waste material and water from the production and manufacture of food to reuse in food production. In addition, the waste products of the astronauts from the consumption of food must be safely recycled for the production of even more food. (A fully closed cycle for food production has so far eluded mankind's grasp.)

Indeed, future space colonists and space travelers might very easily need to be agronomists first and physicist/astronomers second.

At the present stage of technology, hydroponics/plant specialists, microbiologists, and nutritionists would be *de rigeur* on any extended space exploration flight, or on an exploration of nearby planets or asteroids, or for any attempt at a colonization program. Also required would be specialists in the preparation and preservation of the produce and animals grown. Astronauts, working in a very stressful environment, must be kept healthy.

There will be myriad changes in the coming millennium and there are a sufficient number of them associated with or centering on agriculture

6 ■ Food, Consumers, and the Food Industry

A Perspective on Events Leading to the Present

A brief review of the past 1000 years, the developments in the sciences related to food, nutrition and health, and the processing of safe and wholesome food can provide an understanding of how, where, and why the food industry got to "here" in the third millennium. Against this background, an extrapolation to the future can be plotted and problems therein anticipated.

Agriculture has had a long history extending well beyond the 1000 year window of interest here. Where and when man became a farmer and a herdsman is still hotly debated (see, for example, Tannahill, p. 32, 1973). It is not an argument to be resolved here.

At the start of the second millennium, farming, animal husbandry, and fishing were well-established practices. Cooking and baking were finely developed methods of food preparation. Fermentation (here is included preservation by salting as well as the art of brewing) and drying were common means of food preservation to tide people over periods of food scarcity in many parts of the world. There is also clear evidence that food was preserved by techniques that today are commonly called hurdle technology (preserved by a combination of preservative techniques). Such foods as sauerkraut (and other fermented vegetables), anchovies and salted fish pastes, pemmican (sun-dried meat mixed with high acid fruit), smoked, fermented hard sausage and cheeses have a long history of use and are products of multi-component preservative systems. Today's technology can now explain why these products of the craftsman worked as well as they did.

Where did food science and technology begin? The answer to this question will be debated for years. Where the crossover occurred from a craft based on procedures resulting from experiential observations to a technology based on scientific principles will be left to the philosophers of science.

The Past 1000 Years

To provide a backdrop for topics to be discussed in the following chapters, Tables 1.1, 1.2, and 1.3, detail events of the past 1000 years in chronological order. Table 1.1 reviews political events and those voyages of exploration that have shaped history and indirectly had some impact on food. Table 1.2

major historical events that have a bearing on or have shaped those food issues that are yet to be faced in the third millennium. This they are not. They underscore events which I learned (and remembered) from long past history classes, or events that I read about and was impressed by for the influences they have had on scientific and technological development in the agricultural, food, and nutritional arenas.

Table 1.1 An Overview of Important People and Major Developments in Political History and Exploration, 1001 to 2000

Year A.D.	Significant Political Events, Developments or People
1001 to 1100	• Circa 1000: Approximate date for Norsemen's explorations of Greenland and the North American coast, Leif Ericsson in particular, and the establishment of the ill-fated Vinland colony • El Cid (Rodrigo diaz de Vivar) (c.1043–1099): (Took Valencia from the Moors) • 1066: Battle of Hastings won by William of Normandy, feudalism brought to England • 1085: Recapture of Toledo by Christian forces which heralded the collapse of the Muslim Empire in Spain but not the influence of Muslim scholasticism
1101 to 1200	• 1095 to 1291: The Crusades, believed by some to be crudely disguised attempts to extend the commercial power of Italian port cities through the eastern Mediterranean to markets in the Near East • 1137: Rudimentary beginnings of parliamentary government with the *Cortes* in Spain when the Spanish kings began to meet with nobles, the clergy, and members of the business community • 1187: Jerusalem captured by Saladin
1201 to 1300	• 1204: Sack of Constantinople by Venetians with great loss to libraries and universities • Mongol Empire (1206 to last half of 15th century): At its height it united much of Asia from the China Sea to the Mediterranean. The Mongols built roads and canals and encouraged trade with Europe • 1215: Magna Carta signed • 1241: Beginnings of the Hanseatic League which at its

Table 1.1 (Continued) An Overview of Important People and Major Developments in Political History and Exploration, 1001 to 2000

Year A.D.	Significant Political Events, Developments or People
	• 1254 to 1324: Life and travels of Marco Polo, the younger, who for many years was a guest at the court of the Mongols, opened new routes for commerce, introduced new arts, crafts, foods of the Far East (China, India, and Japan)
1301 to 1400	• 1315–1317: Great Famine of Europe brought about by crop failures caused by bad weather and plant disease • 1349–1354: Black Death, the bubonic plague strikes along the Silk Route into and across Europe • Geoffrey Chaucer (c.1345–1400): Best known for the *Canterbury Tales* and who also had a great impact on the English Language
1401 to 1500	• 1431: Joan of Arc burned at the stake • Niccolo Machiavelli (1469–1527): Statesman and political philosopher; author of *The Prince* and *The Art of War* • 1492–1504: The voyages of Columbus and the beginning of exploration and colonization of the Americas
1501 to 1600	• 1517: Martin Luther (1483–1546) posted his 95 theses. This catalyzes the growing desire for reform within the Church and culminates in the Protestant Revolution and accelerates a liberalization of intellectual and scientific thought • 1582: Introduction of Gregorian calendar named after Pope Gregory XIII
1601 to 1700	• 1600: East India Company founded • 1609: Bank of Amsterdam established • 1620: Pilgrim Fathers emigrate to the New World • 1694: Bank of England established. This, with the establishment of the Bank of Amsterdam, broke the monopoly of the old banking families
1701 to 1800	• Benjamin Franklin (1706–1790): American statesman

Table 1.1 (Continued) An Overview of Important People and Major Developments in Political History and Exploration, 1001 to 2000

Year A.D.	Significant Political Events, Developments or People
	• 1787: American Constitution signed
	• 1789: French Revolution, complete break with feudalism with the liberation of the middle classes
	• 1792: Monarchy abolished in France
1801 to 1900	• 1815: Battle of Waterloo, the defeat of Napoleon, and the end of the Napoleonic Wars
	• Florence Nightingale (1820–1910): Nurse whose organization of military hospitals and stress on sanitation greatly reduced military casualties in the Crimean War; made significant improvements in nursing education, sanitary reform, and public health in England and in India
	• 1845: Potato blight devastates Ireland's potato crop
	• 1851: The Great Exhibition opens in London
	• Crimean War (1854–1856)
	• 1865: Abraham Lincoln assassinated
1901 to 2000	• 1914–1918: First World War
	• 1920: League of Nations formed
	• 1939–1945: Second World War; the age of the atom had begun
	• 1945: United Nations formed from which the World Health Organization emerged in 1948
	• 1948: General Agreement on Tariffs and Trade established to work toward removing barriers to international trade in merchandise
	• 1994: World Trade Organization established which replaced the General Agreement on Tariff and Trade. Its purpose is to promote and enforce trade laws and regulations pertaining to merchandise, services, and intellectual property.

Note: Material for Tables 1.1, 1.2, and 1.3 has been gathered from the following sources: Bowle, (1975); Braudel (1981, 1982); Durant (1950, 1957); Forbes and Dijksterhuis (volumes 1 and 2, 1963); Gies and Gies (1994); Itokawa (1976); Morison (1971); Scott (1984); Singer (1954); Tannahill (1989); Toussaint-Sarnat et al.

Table 1.2 An Overview of Some Contributors to Major Developments in Science and Applied Technology, 1001 to 2000

Year A.D.	Event
1001 to 1100	• The Chinese who had developed gunpowder some two centuries earlier now develop firearms • 1093: Use of magnetic needle for navigational purposes known
1101 to 1200	• Craft unions begin to grow in importance in Europe and the trade guilds begin to appear. These associations control the training of artisans, the price of goods, and the quality of products • Leonardo Fibonacci (b 1180): Inspired a rebirth in mathematics. He introduced Arabic algebra and was the first Christian writer discoursing on Hindu numerals, the use of the zero, the decimal notation and the use of symbolic notations for numbers in equations • 1150: The first written record of the distillation of alcohol described in *Magister Salernus* • c. 1170: *Practica chirurgiae*, the earliest surgical treatise in Western Europe written by Roger of Salerno
1201 to 1300	• Paper making introduced into Europe by the Arabs who had in their turn obtained the skill from the Chinese in the 8th century • Albertus Magnus (1193–1280): Beginning of crystallization of the idea that natural phenomena are not to be explained as acts of God but that God acts through natural causes • c. 1280: Convex spectacles developed although these may have had a Chinese origin
1301 to 1400	• Cannon used in Flanders c. 1314 • Cannon mounted on naval vessels (1338) • William of Ockham (? to 1349?): Challenged Aristotelian orthodoxy which began a move to a freer scientific approach. Best known for his famed Occam's razor argument
1401 to 1500	• 1447: Effective application in Europe of movable type

**Table 1.2 (Continued) An Overview of Some Contributors to Major
Developments in Science and Applied Technology, 1001 to 2000**

Year A.D.	Event
	• 1475: Invention of the rifle
	• 1494: Horse-drawn artillery with iron cannon balls mounted on gun carriages appear
1501 to 1600	• Charles de Lécluse (1525–1609), known as Clusius, established first botanical gardens in Europe
	• 1543: Vesalius (1514–1564) published his treatise *On the Fabric of the Human Body*
	• Francis Bacon (1561–1626) moved Western philosophy from its strongly Aristotelian leanings on logic to a more empirical and scientific approach based on data collection and inductive reasoning
	• 1589–1591: First collected edition of Paracelsus's work published. Paracelsus (1493–1541) is considered to have developed the science of chemotherapy
1601 to 1700	• Galileo Galilei (1564–1642) established the basis for Mechanics to be developed as a science with the publication of his *Discourses* in 1638
	• Marcello Malpighi (1628–1694)
	• Anton van Leeuwenhoek (1632–1723) and Robert Hooke (1635–1703): Early classical microscopists whose work bore heavily on later studies on blood circulation, on the nature of sexual generation and on the refutation of spontaneous generation
	• 1660: The Royal Society (London) is established
	• 1666: L'Académie des Sciences established in France
1701 to 1800	• Thomas Newcomen (1663–1729), John Smeaton (1724–1792), James Watt (1736–1819), William Murdock (1754–1839): The work of these men on the design and improvement of the early steam engine provided impetus to the Industrial Revolution
	• 1783: Montgolfier brothers' balloon carries two passengers for the first sustained balloon flight
	• The Industrial Revolution signaled the demise of the

12 ■ Food, Consumers, and the Food Industry

Table 1.2 (Continued) An Overview of Some Contributors to Major Developments in Science and Applied Technology, 1001 to 2000

Year A.D.	Event
	• 1840s: In Hungary, roller milling of wheat was developed which produced a whiter flour with better keeping properties
	• 1846: J. von Liebig published *Chemistry and Its Applications to Agriculture and Physiology*
	• 1858: Reading of a paper by C. Darwin and A. R. Wallace entitled "On the Tendency of Species to Form Varieties; and on the Perpetuation of Varieties and Species by Natural Means of Selection" at the Linnaean Society of London
	• 1859: Darwin's *Origin of Species* published
	• L. Pasteur (1822–1895): Established that fermentation and disease were caused by microorganisms; published germ theory 1861; demonstrated pasteurization of wine and milk in 1865; developed an anthrax vaccine in 1880
	• 1863: National Academy of Science founded in the United States of America
	• 1866: Old and New Worlds connected by transatlantic cable thus inaugurating the beginning of the so-called global village
	• 1866: Gregor Mendel (1822–1884) published his studies on genetics and thus began the study of genetics as a separate discipline
	• 1876: telephone is born as a practical instrument for voice transmission by Alexander Graham Bell
1901 to 2000	• 1901: First transatlantic radio telegraphic transmission
	• 1903: Heavier than air flights begin with the Wright brothers whose machine achieved a speed of 35 mph. By 1970s the Boeing 747 attained a speed of 570 mph. In 1976 the Concorde began passenger service travelling at twice the speed of sound
	• 1915: F. W. Twort and F. d'Hérelle discovered viruses
	• 1919: National Research Council of Canada established

Table 1.2 (Continued) An Overview of Some Contributors to Major Developments in Science and Applied Technology, 1001 to 2000

Year A.D.	Event
	• 1945: The computer age begins; EDVAC (Electronic Discrete Variable Computer) constructed by J. von Neumann; ENIAC (Electronic Numerical Integrator and Computer) built by J. Mauchly and J. P. Eckert. In 1948, the transistor was developed which changed the design of computers and made their use more practical by industry
	• 1957: The launch of Sputnik, the first artificial satellite, marked the beginning of the space age
	• 1961: Yuri Gagarin became the first man in space to orbit the Earth in Vostok 1
	• 1969: N. Armstrong and E. Aldrin, Jr. become the first people on the moon
	• 1971: Interplanetary exploration begins with Mariner 9's voyage to Mars. Viking 1 was the first space probe to launch a lander on Mars which survived the landing (1976)

Others have chosen their own very different lists. Hawthorn (1980) dismisses events and developments from 800 B.C. to 1800 A.D as nothing much more than an accumulation of knowledge or experiences! (But does this not constitute a basis for the beginning of a science and a technology?) Most development, according to Hawthorn, occurred in the (almost) 200 years following 1800. Only six developments, major ones as Hawthorn reports, are worthy of mention:

■ Appert's work on thermal processing
■ Publication of von Liebig's *Chemistry and Its Application to Agriculture and Physiology* (see, for example, Singer, 1954) which signaled the birth of food chemistry
■ Pasteur's studies on fermentation
■ Use of refrigerated cargoes for the transport of perishable foods by Tellier of France and Bell and Coleman of Scotland (for references, see Hawthorn)

14 ◾ Food, Consumers, and the Food Industry

Table 1.3 An Overview of Major Developments in Agriculture, Food Science and Technology, and Food Nutrition, 1001 to 2000

Year A.D.	Event
1001 to 1100	• Improvement made in plough design. The scratch plough suitable only for light soils was replaced by the heavier mouldboard plough of the Slavs • Better crop yields are obtained as the two-field system of rotation is replaced with a three-field system: e.g., wheat in one field, legumes in another, and the third left fallow • Horses begin to replace oxen in farming operations in Europe with consequent change in grain-growing practices • Cider, perry, and beer (usually fruit- or spice-flavoured) were common beverages. Hops began to be used to flavour beers. Wine is also a common beverage • 1086: Domesday Book recorded presence of 5,624 mills (largely water mills) in England
1101 to 1200	• Muslim conquests bring new foods and farming practices including irrigation of crops to southern Europe from Middle and Far East. Crusades bring greater familiarity of Muslim food culture and habits to Europe
1201 to 1300	• The Hanseatic League banded together merchants within its member cities for both physical and economic protection. The League served to control the price of raw materials and the quality and price of finished goods including foodstuffs. It also could establish boycotts and courts of dispute regarding commerce and labor. • 1276: In Ireland, the first official whisky distillery was established
1301 to 1400	• C. 1330: Willem Beukelszoon introduced the gutting of herring which resulted in a better quality salted product able to last longer • 1330: Street cleaning regulations were enforced in Göttingen, Germany

Table 1.3 (Continued) An Overview of Major Developments in Agriculture, Food Science and Technology, and Food Nutrition, 1001 to 2000

Year A.D.	Event
	• 1369: Edward III notified officials of City of London of having received complaints from residents near the slaughterhouse of St. Nicholas that the butchers slaughtered animals in the street and carried the waste to the Thames into which the waste was thrown
1401 to 1500	• 1474: *de Honesta Voluptate ac Valetudine* written by Platina, librarian to the Vatican, was published. This was said by some to signal the start of French haute cuisine
	• 1482: Ordinances for feeding standards were decreed for craftsmen in Germany. These described the number of courses per mid-day and evening meals on meat days and meatless days
	• 1497 or earlier: European fishermen, particularly Basques, fished off the east coast of Labrador, Canada
1501 to 1600	• Introduction of many new fruits and vegetables to Europe as the voyages to the Americas progress
	• 1518: At the Diet of Innsbruck it was noted that importers added brick dust to ginger and mixed unhealthy extraneous material with their pepper
	• "The German peasant feeds on brown bread, boiled beans or peas, drinking water or whey:" Written by the German humanitarian, Johannes Boëmus, in 1520
	• Hugh Platt (1552–1611): Investigated the use of salt and marl in agriculture
	• 1599: East India Company formed by English merchants annoyed with Portuguese at the rise in price of pepper from 3 to 8 shillings a pound
1601 to 1700	• 1609: Tea introduced into Europe and becomes very popular by the end of century
	• 1684: Earliest records of beer consumption in England. By 1689 beer consumption had reached 832 pints per person per year or 2.3 pints per person per day
1701 to 1800	• James Lind (1716–1794): Lind, familiarly known as the

16 ■ Food, Consumers, and the Food Industry

Table 1.3 (Continued) An Overview of Major Developments in Agriculture, Food Science and Technology, and Food Nutrition, 1001 to 2000

Year A.D.	Event
1801 to 1900	• 1812: N. Appert's work *The Art of Preserving All Kinds of Animal and Vegetable Substances for Several Years* published in English
	• 1820: F. Accum's book *A Treatise on Adulterations of Food and Methods of Detecting Them* was published
	• Kanehiro Takaki (1849–1920): Reported on the cause of beriberi by establishing food as a factor and defined prophylaxis of beriberi
	• 1853: The first canned condensed milk produced by Gail Borden
	• 1859: Great Atlantic and Pacific Tea Company was created signaling the beginning of the food chain store era
	• 1860: Food and Drugs Act passed in Great Britain
	• 1862: Morrill Act established the landgrant college system in the U.S. which greatly influenced research into agriculture and food processing
	• 1862: Reliable machinery was available for machine-made ice
	• 1868: The food company Libby, McNeill, and Libby was established in Chicago
	• 1869: Mège Mouriès was granted English and French patents for his invention of margarine
	• 1877: The ship, the Paraguay, delivered the first frozen beef from Argentina to Nantes, France
	• 1882: Canned Foods Exchange was incorporated and is the oldest canner association in the world
	• 1885: *Good Housekeeping Magazine* was first published
1901 to 2000	• 1903: The first "sanitary" can produced
	• 1906: G. Hopkins published first report on vitamins (reported in Hawthorn, 1980)
	• 1927: The first vitaminized margarine was marketed
	• 1929: Clarence Birdseye developed technique of quick freezing of food products

Table 1.3 (Continued) An Overview of Major Developments in Agriculture, Food Science and Technology, and Food Nutrition, 1001 to 2000

Year A.D.	Event
	• 1958: The constituent assembly of the European Council of the Codex Alimentarius was held in Vienna
	• 1958: Delaney clause added to the Food Additives Amendment
	• 1962: Rachel Carson (1907–1964) wrote *Silent Spring* which crystallized worldwide concern for the environment
	• 1962: John F. Kennedy enunciated the basic rights of consumers, i.e., the right to safety, the right to be informed, the right to choose, and the right to be heard
	• 1964: The Institute of Food Science & Technology of the UK (IFST(UK)) founded
	• 1970: International Union of Food Science and Technology was inaugurated in Washington
	• c. 1975: Development of principles of intermediate moisture foods (see, for example, Karel, 1976)
	• 1976: Development of the concept of hurdle technology for the preservation of foods (Leistner and Rödel, 1976)

Meanwhile, Goldblith (1989), writing on the occasion of the Institute of Food Technologists' 50th anniversary, is much more generous. He describes a number of broadly significant, far-ranging developments in food science and technology in the 50 years of the Institute's existence. Several but not all those developments mentioned by Goldblith in this narrow time span are noted in the tables.

The tables, despite their singular and personal nature, provide pegs on which to hang particular events in order to see them collectively. They telescope one's view of past, significant happenings to be discussed later. They are the progenitors of things to come.

It is best to look at these tables of activities of the past 1000 years synoptically. This allows a quick integrated overview — and that is all that is needed — at where one has been, what happened then and how other activities, particularly food-related events, were influenced overtly

18 ■ Food, Consumers, and the Food Industry

Political and Historical Developments and Voyages of Exploration

The Crusades spanned a period of some 200 years. They were an attempt to free the Holy Lands from Islamic control although some historians have viewed them as an exercise by the Italian and Venetian city states to gain control of the lucrative trade routes to the Middle and Far East. Spices were an important commodity, among many others, that were traded along these routes.

The Crusades had two effects for the purposes of the story here. First, they were an economic drain on the manorial system of agriculture which was characteristic of the feudalism that earlier had swept Europe. They wasted manpower that was needed on the manorial lands by carting such resources off to war. They also wasted leadership: The lords of the manors and their stewards were busy fighting in foreign lands.

Second, they also brought new foods, cuisines, and technologies from the Middle and Far East to Europe. There is little history to tell about who the early agronomists were who first experimented with the cultivation of these crops in the Old World or adapted them to the European diet.

Those innovations that were attractive or might have been seen as profitable were brought back probably by the stewards accompanying their manorial lords. They knew the land and saw an opportunity to exploit the use of the new crops. Foods, if found acceptable, often become staples in the diets of their discoverers supplanting the indigenous plants. They then become a new cash crop. From these new homes by various routes they are disseminated far and wide in their new lands. But this is conjecture. The result was that the Crusades introduced new foods into Europe and presaged the end of the manorial system of agriculture.

The Black Death raged across Europe periodically. Historians have estimated it to have killed one in three persons. This disease continued the loss of manpower started by the Crusades and represented a huge economic loss. Labor, especially farm labor, was decimated.

Wages were rising in the towns as industries began to grow in economic importance. People flocked to the larger towns and cities where the higher wages were and forsook their feudal obligations. This migration of workers from the fields to the towns was one more nail of many in the coffin of the feudal agricultural system.

In a different vein, there was also a beginning of liberalization of intellectual, artistic, and scientific thought. The peg for this was chosen as

Marco Polo, Luther is the one having the largest popular following and providing the best known peg to trigger the imagination.

In any event, Europe, now the center of world attention, was in a turmoil.

Developments in Science and Technology

As might be expected, historians of the era differ somewhat in their opinions of which developments were the definitive ones, the ones that shaped the face of technology. Gies and Gies (1994), for example, give greatest import to

> The advent of printing and the earlier development of paper, a technological import from China
> Gunpowder, cannons, and small arms and their impact on warfare and civilization
> The development of nation states

Certainly each development brought huge changes and had a domino-like effect on the social and economic structure of the then civilized world. Indirectly they also had a great impact on food and agriculture.

The importance of gunpowder and weaponry is given similar credence by Braudel (1981) but placed in a different order in the hierarchy. He also sees greatest value in the development of paper and the printing press. Next in importance, he places improvements of transport which he saw as improving communication and trade and unifying geographic areas.

Some applications of technology had unexpected impact in areas remote from their immediate intended applications, for example, in food and agriculture. The following sequence of events is woven together rather simplistically from several sources but especially from Braudel (1981) and Gies and Gies (1994):

- Weaponry utilizing gunpowder dramatically changed warfare. First the use of cannon in warfare put an end to the defensive advantages of high-walled castles, walled cities and towns.
- Cities and towns, in their turn, turned to the use of such weaponry for their defense. But this sophisticated weaponry required a skilled, professional soldiery. The defense of cities and towns,

20 ■ Food, Consumers, and the Food Industry

- The design of such weapons was vastly improved. They became more accurate and mobile. Improvements made gunpowder more safe and reliable and hence more useful. Handguns, by 1425, became one-man weapons.
- The armored knight, often a member of the feudal nobility, became obsolete. An infantry armed with musketry was cheaper to arm, much more mobile, and had greater killing power. The feudal nobility lost much of its political influence and power.
- Gunpowder and cannonry as well as small arms and standing armies were expensive to maintain. An entity such as a noble, or even a village or town found such defenses beyond its financial means. They allied themselves, therefore, to other larger cities. Only a central government could afford the high cost involved in outfitting such an army.
- Nation states, thus, came into being with central governments charged with raising taxes to pay for defense. Eventually this, in turn, led to other infrastructure costs for services such as sanitation, paving of roads, and the building of massive national projects; for example, the reclaiming of land from the seas as in The Netherlands or the building of cathedrals.

As cities and towns grew, so did the craft industries needed in them. Agriculture was still the main industry but, much to the concern of the large landowners, the crafts, weaving, glass work, carpentry, arms and the armorer, and the smithy grew in importance. The armorer turned smithy became a gunsmith. Cannons required a foundry and iron workers were needed but so, too, were miners for iron ore and coal for smelting. Bells for the great cathedrals that were being constructed also needed iron and also required their special craftspeople to tune them properly.

The growing demand for iron ore and coal for the foundries brought the need for deeper and larger mines. With this came the need for ancillary equipment like pumps to keep water out of the shafts and to keep air flowing down to the miners and lifts to haul the miners and the ore to the surface. Mining engineering developed.

Paper and the development of the printing press brought their own changes. Paper was much cheaper than parchment and lent itself more readily to printing since it had a smoother, more even surface. Flax and hemp cultivation had lead to the manufacture of linen which in large

How did this influence both agriculture and food? What appears in the modern day as a simple, indeed logical in hindsight, progression as mounting cannon on wheels brought great changes in agriculture. Improvements in the horse collar saw oxen give place to the horse. After all, the horse could pull both the plough and the mobile cannon and also be an instrument of war. The horse was far more mobile than oxen and less troublesome. Oats, therefore, became an important cereal crop and areas where oats could be grown became economically successful.

Then, there was the growth of a non-agricultural class of people who depended on farmers for their food but themselves were not farmers. These were the tradespersons and skilled craftspersons as well as the clerks and sales personnel and laborers required to keep the industries going. Specialization was developing and a commercial food industry grew up with such artisans as brewers, bakers, and confectioners to be able to feed these consuming individuals.

With the skilled crafts came the guilds, organizations of individual craftspersons in the 12th century. They began to control their crafts, develop standards, regulate prices, and set wages. Europe, in particular, was moving from becoming a developing area to being a developed one. It was changing from a farming economy to an industrial economy.

The establishment of scientific societies allowed the transmission of scientific developments to a wider audience. Scientists then communicated with one another through lectures, the transactions of which were written up, and by letter. Today such communication is done by lectures at scientific conferences, or between scientists working in similar fields who are in constant contact by e-mail or by publication in some scientific journal.

Now scientists have reverted to what they did 200 and 300 years ago. They are writing to colleagues but now the vehicle for communication is e-mail. It is cheaper, faster, and response can be immediate. Indeed, scientific journals as a means of disseminating scientific knowledge may soon reach the end of their usefulness (Strauss, 1996). The next step could be the electronic journal. Strauss comments that electronic journals if put out by scientific societies could be 'published' at a fraction of the cost of the print journals. Print journals are becoming so expensive that many universities are having to cut subscriptions (Strauss, 1996). Such action is hardly conducive to the communication of science.

The advent of nation states meant that roads and canals were built or improved. Improved travel led to better communication with the devel-

Breeding techniques were known even in the earliest years of this past millennium. Horses, cattle, dogs, and even fish were certainly bred for desirable traits. Mendels' work (see Table 1.2 and Singer, 1954) published in 1866 (largely ignored until 1900) introduced a scientific basis for understanding observations in cross-breeding programs but it was not until genetic manipulation became a fact that real improvements could be made to the food supply.

Developments in the Agricultural, Food, and Nutritional Sciences

Food trades were specialized and organized well before the 1000-year review detailed here. Associations of specialized artisans such as bakers, brewers, and winemakers go back to ancient Roman times. In the 11th century reference is made to an association of victualers. Since many households in earlier centuries did not possess ovens for cooking, house-holders took their bread to the local baker for baking or purchased their meats already roasted from public cookshops, some of which became very famous in England.

A good overview of the state of agriculture can be had from the *Domesday Book*. Its creation was decreed by William the Conqueror in 1085 and was finished in 1086. It provides at the beginning of the second millennium an excellent catalogue of what was available in England respecting agriculture (see, for example, pages 463 et seq. Maitland, 1897). Maitland reviews the statistics in the *Domesday Book* and uses it to describe the agricultural potential of the many manors in the various regions of England:

- How the manor is rated according to a certain number of 'units' it contained
- That the manor contains land for a certain number of oxen or teams of oxen
- That some of the teams or oxen belonged to the lord and some to the men

From this data and with the help of some population statistics Maitland prepares an almost bewildering array of agricultural statistics. Efforts have been made to estimate what yields might have been, but these are largely

The *Domesday Book* recorded the presence of over 5000 mills in England (Gies and Gies, 1994). The only purpose for a mill is for the milling of grain for flour; flour would only be used in breadmaking. Since records of a century earlier than the *Domesday Book* enumerated only 100 mills, such a profusion of mills indicates clearly that there had been a shift in diet. This shift appears on the evidence to be from a diet consisting of a boiled coarse porridge to one that indicated a greater consumption of baked bread.

Fruit- and spice-flavored beers very common in the earlier centuries of the millennium were gradually replaced by hopped beers. By the 14th and 15th centuries hopped beers had surpassed in popularity the fruit- and herb-flavored beers. There also were improvements in the quality of salted herrings.

The greatest areas of achievement were perhaps in agriculture. In China there is clear evidence of the development of new strains of rice during the Sung Dynasty (Freeman, 1977). New foods were introduced with the Moorish conquests across North Africa and into Spain. Then with the voyages of discovery more new foods were brought to Europe. The voyages of Columbus triggered an exchange of foods from the Americas as had the Crusades and the later travels of Marco Polo triggered interest in the foods (and culture) of the Middle and Far East in earlier eras.

Some examples of this dissemination of food varieties can help understanding here. Peppers, the *Capsicum* variety, can be traced rather precisely from the New World to Spain where they were brought by Columbus. Peppers were certainly in Spain in 1493. They are recorded in Italy in 1526, in India in 1542, in Germany in 1543, into the Balkans in 1560, and into Czechoslovakia in 1585 (Andrews, 1990). From here they were moved into the Far East. As *Capsicum* peppers moved eastward they became known as paprika.

Tomatoes are also a New World introduction into Europe but one which had a much slower acceptance. They come back to the Americas by a circuitous route. Pier Andrea Mattioli in the 1554 edition of his herbal recorded the presence of tomato plantations in Europe. He further recorded that tomatoes were eaten in Italy with oil, salt, and pepper, a custom unchanged to this day (Rick, 1978). Several old herbals and herbaria contained colored drawings of tomatoes (Brücher, 1983). By the end of the 17th century tomatoes as well as capsicum peppers enlivened the meals of Europeans.

24 ■ Food, Consumers, and the Food Industry

The major commercially important food crops and animals to come from the Americas were (Goldblith, 1992; Mermelstein, 1992):

chili peppers	potatoes	tomato	beans
peanuts	maize	sweet potato	cassava
pineapple	guava	avocado	papaya
cacao	squashes	turkeys	chicle
Muscovy duck			

Other crops such as quinoa and amaranth did not have an instant success but now are gaining in popularity and in commercial value in world markets.

In the true sense of exchange, the Americas received from the Old World a wide variety of food crops and animals (Mermelstein, 1992):

wheat	barley	oats	rye
rice	lentils	chickpeas	sugarcane
olives	tea	yams	taro
oranges	limes	lemons	apples
pears	cherries	plums	walnuts
almonds	pistachio	hazelnuts	black pepper
cloves	nutmeg (mace)	cinnamon	ginger
pigs	goats	sheep	chickens
geese (some)	ducks (some)		

The Old World, or Eurasian, contribution to the foods of the New World seems rather immense. It must be remembered, however, that the Eurasian contribution to the variety of foodstuffs is itself a fusion of northern European, Mediterranean, Middle Eastern, Far Eastern, and African agricultural practices.

Africa made its contributions both to the Americas and to the cuisine of Eurasia. To the Americas the contribution was in the form of such crops (Mermelstein, 1992) as:

sorghum	yams	sesame
okra	bananas	plantain
watermelon	rice	coffee

The greater variety of foods that became available also caused major changes in agricultural practices. The grape took hold in France, Spain, and Italy with the result that these are wine producing countries; wine is the preferred drink. In the colder North where wheat predominated, beer was the preferred tipple. Also in the south the olive grew well and became a profitable cash crop. Its oil was useful in cooking and its meat proved a valuable resource as animal feed.

The beginnings of the science of nutrition arose beginning from the middle of the 18th century onward when there was a real understanding of the value of food as a tool to combat certain diseases. Much of this understanding was the direct result of the early work by Lind (Singer, 1954), Takaki (Itokawa, 1976), and Hopkins (Hawthorn, 1980).

Appert's work on the development of thermal processing of foods for preservation had several interesting outcomes for the food industry and for the diet of people in general and for exploration and warfare in particular:

- It began a break with the seasonality of food and its dominance over people's diets. No longer was meat only plentiful at slaughter time in the fall when the herd was culled in proportion to the amount of feed available to last the winter. Salt meat was not the only method to conserve meat. Fruits and vegetables could also be conserved for use over the winter months.
- Food could now be stockpiled by governments for use in emergencies or periods of scarcity. Ships were now able to engage in longer voyages without having to rely on their proximity to landfalls in order to replenish their supplies.
- A greater variety of food was available to all classes of people.

The canned food industry had begun.

With Appert, Accum, and von Liebig, scientists were beginning to apply their training to study food and its chemistry as a separate branch of science to be researched for its intrinsic value. Problems associated with foods and their purity were publicized and so began the long road to a safer, purer food supply. (Accum was not a popular person with food manufacturers because of his analytical findings on the purity of food. He was forced to flee England because of their wrath.) Despite these early workers, the emergence of the food sciences as a separate and distinct

safety, quality, and new product development (Holmes, 1996). In the U.S., the establishment of the landgrant colleges certainly boosted the scientific effort put into the sciences applied to food and its processing.

What Did the Common People Eat?

Trying to learn what the dietary habits of different people were through the second millennium and to compare them on a time basis are confounded by ethnic, cultural, social, and economic factors. The observations are clouded by geography, which has environmental and climatic factors, and the different agricultural practices adopted by cultural traditions and the acceptance, readily or not, of new crops, fruits, and vegetables.

Any conclusions regarding diets can be very confusing for many reasons. First, there were very distinct class differences. These distinctions were based on both economic and social status throughout the early centuries of this past millennium. The rich, generally defined as the nobility but also including the merchant class, could afford to eat well. Certainly by the standards of the early centuries, the rich ate better than their villeins but by today's standards it was not attractive fare in either instance. Farm workers did not eat as well but even here opinions differ; some claim the diets of the rich were fatty, very rich, low in vegetables, and hence nutritionally inferior while the diets of the poor were better nutritionally since they were high in cereals and vegetables with occasional meat or fish to supplement their diets.

Second, much of the evidence for what people ate comes from early literature that describes foods which were eaten at feasts or at religious, festive, or celebratory events. It is, therefore, not indicative of what the rich ate routinely and certainly not what the poor ate. Even when they ate at the lord of the manor's castle, they received the bread which the upper classes had used to sop up the gravies and sauces. It was considered a discourtesy to those having it and a display of very bad manners on the part of the upper class to bite into this bread (Hartley, 1954).

An investigation of the meals served on grand occasions can be very revealing of the skill of the cooks and the great variety of food that was available. People ate a much wider variety of foods than is consumed today. However, there is little true indication of the regular diet of people from these early cookbooks.

They are interesting in that in their transformation to modern culinary form they are not far removed from those found in cookbooks of today. The main exception might be that today's consumer would like a less fatty diet. Cosman (1976) in a much more learned treatise on English cookery covering this same Medieval period describes not only recipes, again in translation to modern ingredients, but also describes menus, customs, food ingredients, market laws, and the foods eaten by the various social classes; de la Falaise (1973) discusses and describes recipes from the 14th century to the present day.

This same millennial period is described by Hartley (1954). Her book, however, provides a much more complete description of the medieval kitchen, its utensils, techniques for the preserving of food, the agricultural practices during the centuries, and the foods brought to the table. It touches heavily on what one would call hearty country cooking. Instructions on how to behave at table can be found here along with excerpts from housekeeping diaries which describe foods and obligations of the better classes toward the poor.

One gets some feeling for the food available for many of the English people. On the other hand, one is always confronted with doubt about how representative these recipes, primarily from English sources, are for the rest of Europe and for what cross-section of the population are they representative. Nevertheless these books cover the entire gamut of recipes for English cooking for the past millennium.

Thus one can see the shift in diets that has occurred. For the wealthy during the past millennium, meat, fish, and game were available and people ate well but certainly not in any healthy manner that follows the nutritional guidelines of today. For the poor, the situation was very different. Cereals and vegetables made up a large part of their diet and were supplemented by cheap, fatty cuts of meat when they were available.

Some awareness of the extent of nutritional knowledge extant in 1828 can be had by examining recipes for "sick cookery" and "cookery for the poor" (A Lady, 1828). For the poor, some brewis is recommended:

> *"Cut a very thick upper crust of bread, and put it into the pot where salt beef is boiling and near ready; it will attract some of the fat, and when swelled out, will be no unpalatable dish to those who rarely taste meat."*

Or, in another section, the cook is enjoined to save bones from the family's

wine with sugar, nutmeg, and lemon peel with bread crumbs stirred in) or caudle (a fine smooth gruel of half-grits, boiled and strained, to which sugar, wine, lemon peel, and nutmeg can be added with a bit of brandy).

A middle class gentleman's diary, the diary of a parson, provides an interesting perspective of life, and food, in the latter half of the 18th century (Woodforde, 1985). There are no recipes but descriptions of meals eaten in his home or taken as he visited his charges around the countryside of Norfolk, England are described.

Outside of festive occasions the daily food for all classes of people by present day standards was dull and uninteresting and, for the most part, lacked variety. Geography, season, and climate dictated a large part of their diet. Fresh produce was only had in season and fresh meat served frequently only at slaughtering time in the fall.

In Colonial North America

This bland, monotonous fare of the Old World was certainly a fact of life in North American colonial times. Food habits, of course, varied widely from north to south. In the south, crops could be grown year-round and animals could be kept over winter months. In the north, it was necessary to preserve food for survival over the winter months. The usual preservation techniques were much the same as those used in Europe:

- Cold storage for vegetable crops in root cellars
- Dehydration of fruits and fish and meat; salting or corning (often in combination with drying) of meat (hams and beef) and fish
- Fruit preserves made with sugar
- Fermentation or pickling of fruits and vegetables (acidification)
- Preparation of fruit beers, ciders, and wines
- In a primitive form of canning, preservation in crocks sealed with a layer of fat to exclude air and, unbeknownst to them, bacteria

The latter was certainly not performed on any scientific basis. Like other preservative techniques of the day, it was probably the result of long observation of trial and error methods.

The Institute of Food Technologists published (Anon., 1976) a booklet describing food in the colonial days from roughly the 1700s. Here the techniques of preserving food were listed as cold storage, drying and

colonists pigs were a better animal than sheep to keep. First, pigs were not fussy eaters as were sheep and they could forage for themselves. Second, their litters were up to ten or more piglets while sheep gave rise to one or at most two lambs. Consequently, North Americans are traditionally more accepting of pork than lamb.

In a diary kept by Mrs. Simcoe, wife of the Governor of Upper Canada, life in Upper Canada (basically represented by present day Ontario) is described in the period 1791 to 1796 (Innis, 1965). Here all the vicissitudes of colonial life are described and in the descriptions can be found ample references to the foods that were available not only to the upper classes but also to the travelling public as Mrs. Simcoe accompanied her husband on his travels in the performance of his public duties.

Another look into colonial life in the New World can be found in writings of Catherine Parr Traill (1855). Traill originally described her book as a manual for Canadian housewifery. Foods of the early settlers are described as well as the general procedures for preparing them for meals and for preserving them. However, in addition, she provides common-sense insight into some of the reasons for the food habits. For example, when discussing beer Traill comments that the settlers lament the lack of good beer and ale. If there were better beer, "there would be a thousandfold less whiskey drunk in this colony!" But she explains that the lack of good beer is due to the scarcity of maltsters and that barley is not grown as a rotation crop with which to make the malt. In addition there was a lack of large containers in which to make the beer.

Food in colonial America in the Connecticut valley is well described in a booklet put out by the Women's Alliance of the First Church of Deerfield published first in 1805 (Anon., 1897). Food, for what was mainly an agricultural way of life, was hearty. Breakfast for farm hands consisted of sausages (often with fried sweet apples) or ham or souse (pickled pig's feet, ears, and skin), fried pork and eggs, or boiled freshened salt mackerel or shad with boiled potatoes. This would be accompanied by Johnny cakes, or hoe cake, rye or Indian bread, flour bread nut cakes or pie. Dinner was often boiled corned beef, pork, pudding, and whatever vegetables were in season (turnips, cabbages, pumpkin, or squashes). Supper was cold meat, leftover vegetables, brown bread, or baked sweet apples.

In China

The Sung Dynasty saw some very important changes occur in agriculture (Freeman, 1977). First, new strains of rice appeared – early evidence of the ability of the Chinese to breed better crops — which matured earlier and gave better yields. Their use by farmers permitted double-cropping. Needless to say the use of the new strains was encouraged by the government since it was they who taxed on the amount of rice produced. Although there had never been broadly based famine in China, food abundance now far outstripped the needs of the population.

Second, agricultural commercialization began as merchants contracted for crops, paying higher prices for the better quality crops. It is here in China that the first evidence for such agronomic practices that are common today can be seen. Sugar cane, litchis, and tea became cash crops. Tea production, however, was carefully controlled by government.

Markets contained meats and fish of all kinds and a tremendous variety of fruits and vegetables. Freeman records over 30 varieties of vegetables, with 17 varieties of beans alone.

There were class distinctions as there were in Europe. Eating habits reflected the customer/consumer's wealth and social position but the methods of cooking and for the most part, the ingredients, were the same for all social classes. The major differences between the classes were in the greater proportion of poorer quality grains, vegetables, and meat used by the poor than by the wealthy. The wealthy were able to afford better quality foods and the better cuts of meat.

A food service industry flourished in China much earlier than in Europe. In the cities there was an abundance of restaurants (often hired to cater banquets and special occasions). There were theme restaurants (vegetarian, beef only or mutton only, etc.) and as well wine- and tea-houses and noodle shops abounded. So, fast-food restaurants are not a recent innovation. They have been common in China for several hundreds of years. Street vendors dispensing what today is called finger food were as common in Chinese cities centuries ago as they are becoming common in many European and North American cities now (Freeman, 1977; Mote, 1977).

With their penchant for fresh food, the Chinese developed agricultural practices for growing vegetables well into the winter months. By covering plants with straw matting and by using the heat of rotting manure dug into deep beds to warm the soil above, they were able to grow vegetables and prevent frost kill (Mote, 1977). They also bred hardy varieties of vegetables that resisted the cold.

service and convenience. About 35 years later, *Good Housekeeping Magazine* started publication in North America. Such publications brought recipes, nutrition information, cooking and household hints; more importantly, they created desires for new food adventures in the housewives and mothers who were the dominant customers. The emergence of the customer as a force in food retailing began.

In North America this growth of the consumer accelerated with the publication of Rachel Carson's *Silent Spring* and the emergence of consumerism and the activities of Ralph Nader, a consumer advocate. There then began a questioning by the customer/consumer of the veracity of government, big business, and science and concomitantly express concern for the environment and the safety of the food supply. This concern for food safety began slowly early in the millennium by Edward III, then by the Diet of Innsbruck, and later by Accum who so raised the ire of food manufacturers he had to flee England. The stage was set for a polarization of goals, attitudes, and traditions respecting food.

The Present Status

Historically, most of the peoples in the world have lived marginally. They primarily ate cereals, root crops, and other starchy foods (e.g., potatoes, bananas) supplemented by meat or fish. There were few recorded instances of mass starvation in the second millennium. Those instances of starvation that have occurred usually have been the result of

- Natural crises such as flooding, or infestations (the Irish Potato Famine), or drought which cause losses of regional harvests of critical commodities
- Mismanagement of natural resources through pollution of water or desertification of land, rendering it unfit for farming
- Lack of an infrastructure and resources to prevent post-harvest losses or to distribute food
- Warfare and the deliberate devastation of food resources to cripple opponents

A sharp distinction must be made here between starvation and malnutrition. Famine resulting in starvation, except in the situations mentioned

habits. People do not have the resources to buy sufficient food and are ignorant of the proper foods to buy with whatever limited resources they do have.

The other aspect of malnutrition is best described as overnutrition, i.e., obesity. This brings with it the same infirmities as malnutrition caused by poverty and ignorance of good nutrition and irresponsible (through ignorance) food purchasing. Thus malnutrition can be seen to have two causes: poverty and ignorance. The theme of the general public's ignorance of food, its production, its safe and proper handling, its preparation, and its physiological and nutritional importance will recur throughout this book. The understanding of the public will be essential to the resolution of the many problems that will be faced in the third millennium.

Population Growth

Estimating today's population is relatively simple. Problems arise when demographers attempt to calculate what populations were in the past so that they can understand factors in population dynamics. They also want to know the rate of growth in order to make predictions of what the population will be in the future. Planners and policy makers can determine health, nutrition, and housing needs for future populations. Knowing how quickly population is growing allows them to anticipate future needs.

Population growth can present major problems. Available food and population are closely related and this is important data for governments to consider in formulating food policy. The Chinese were among the first to recognize this when the Emperor Tai Tsu decreed in 1368 that a register, the "yellow register," be created every 10 years recording the number of mouths to be fed in each household. This was placed at the gate to each household (Broadbent, 1978). Continued unfettered growth means more demands on land resources and the oceans to keep the world fed and housed. Land, especially agricultural land, and the waters of the world are finite resources for food production, but there seems to be infinite potential for population.

Braudel (1981) discusses the difficulties in getting any real valid historical data for calculating world population dynamics. Therefore, without good data, it is impossible to extrapolate from what the world population was to what it is today and hence forward to what it might be projected. The difficulty, as always, is that real figures do not exist. There are some

are always introduced. For example, enumerations were generally carried out for taxation purposes but who were counted? Were women? Children? Slaves?

A simple projection based on data in Braudel (1981, page 42) shows that present world population of approximately four billion would grow to be between 6 and 7 billion by the year 2050 and to between 20 and 30 billion by 2150. Another estimate (Anon., 1998b) put the present population at 5 billion (no reference as to source of statistics provided) and predicted a doubling in another 50 years at the present rate of growth.

The United Nations has covered itself by hedging on the number of children per female for its predictions for population growth (Anon., 1998c). By the year 2150, it projected a low population estimate of 3.6 billion, a medium estimate of 10.8 billion by the same year or a high estimate of 27 billion. While there are many factors which can influence the discrepancies in these projections, the principle factor is the number of children per female. The implication is clear: Birth control is the main method to slow population growth.

No matter what assumptions population estimates are based upon, it is obvious that feeding future populations will become increasingly difficult as the third millennium begins.

Hulse (1999) discusses the major problem that population growth is spawning: urbanisation. People move to the cities from the rural areas with the consequence that non-farm workers far outnumber the farm workers. The proportion of the North American and British work force employed on farms compared to those in non-farm employment is less than 1 in 20 and has been steadily dropping for the past 100 years (Hulse, 1999). This is happening in the less developed as well as in the developed countries. In 1970, urban populations in developed countries as a percentage of the total population was 68% and this is projected to grow by 2010 to 79%. The corresponding figures for less developed countries for the same period is a growth of the urban population as a percentage of the total of 25 to 49% (Hulse).

A migration such as this has obvious implications for the agricultural sector, for example:

■ Rural populations must become more efficient to supply the demand for food but at the same time they become more productive, rural employment opportunities are reduced. That is, increased quantities of many crops cannot feed large numbers of urbanites; for example, wheat and corn will predominate over other grains such as millet, sorghum, oats, and quinoa.

■ Urban populations are more discriminating in the quality of food they buy. They demand produce of uniform ripeness and demand that their produce be of uniform shape and size. They (food manufacturers in particular) also require that produce have uniform functional properties.

These are difficult challenges. Urban population growth has ties to the food microcosm, particularly the agricultural end, which are not clearly understood. Urban food security is a neglected area of knowledge (Hulse, 1999).

Food Consumption: New Dimensions and New Problems

There is hunger, indeed starvation, in many parts of the world, even in those areas where there is plenty. Yet in other parts of the world, there is an excess of food. News reports describe farmers who dump food because they are angry at the prices they receive for their produce or are unhappy with government policies respecting food imports. Food is obviously very unevenly distributed throughout the world.

Only partly can influences on food production such as soil, climate and available arable land account for such global variations in food distribution. Many factors influence who has and who hasn't enough available food to eat, and this must be resolved before the deficiencies in distribution can be corrected:

■ Non-availability of birth control policies and education in birth control for women in some countries is often hand-in-glove with high population densities, with poverty, and with ownership of available arable land. Cultural practices and religious beliefs in these areas frequently oppose birth control education.

■ Cultural or religious practices which have deemed that certain foods are acceptable and others are unacceptable. Examples abound: the Jewish or Muslim practice of refraining from eating pork or the Hindu non-acceptance of beef; eating horse or dog meat is a taboo in many cultures; and North Americans as well as

- Economic factors play an underlying role in food distribution. Price and pricing policies for food determine what farmers grow and what price customers must pay for food. Where there is poverty even amid plenty, the poor cannot afford food. Where government policy dictates a desire for export sales, farmers are moving to the production of non-food cash crops at the expense of their indigenous crops. Crops for export (sugar, cotton, coffee, tobacco, etc.) bring needed cash.

One trend that is growing rapidly is a desire among consumers of the developing world for meat. Meat provides a unique and desired flavour to otherwise bland vegetarian dishes and as economies of nations progress their peoples want meat. Hulse (1999) reported anecdotally that meat consumption in the past decade has grown by nearly 15% in Indian cities.

This desire for more meat products does raise some dilemmas. It does throw into conflict meat-eaters against vegetarians. The consequences of such conflict can be trivial or serious as different cultures or religions or animal rights groups clash over the issue of meat eating or raising animals for food.

The argument is often pushed further by the advocates of vegetarianism when they argue:

- That large areas of land which could be used for cereals or grains for people must be devoted to raising and feeding grazing animals;
- That only the wealthy can afford meat products since meat must always be more expensive than cereals on a weight for weight basis; and
- That a greater weight of grains suitable as food for people must be fed to animals to produce a lesser weight yield of meat.

A shift in power to the food consumer from the primary producer as the key element driving the food chain has unfolded in the past millennium. The consumer has become a powerful driving force; food manufacturers and retailers alike have been motivated to cater to the whims of the consumer.

Technology has played a major part in this power shift. Techniques for preserving food have allowed an enormous variety of safe, nutritious food to be available year round. Ingredient technology has allowed the

The consequence of all this technology has given freedom to customers and consumers in developed countries to choose from a wide variety of fresh and processed foods all at a reasonable price. In addition, customers and consumers alike have many options for places where they may purchase their food or where they may eat their food. And each choice of food, however it was prepared, or each eating location, represents or was created by an advance in technology in food production, processing, distribution, marketing or retailing, and advertising. The food consumer, therefore, has become somewhat empowered.

But has this empowerment come with a price? There are developments which might suggest that neither the customers nor the consumers are really empowered but merely think they are getting what slick advertising tells them they want or need so well researched are their buying habits and motivations.

Has convenience gone too far? By catering to the whims of the customer and the consumer, technology has developed products that permit the cook, for one example, greater freedom from the drudgery of food preparation. Many now eat out. Or they can buy semi-prepared foods that simply require putting components together. Statistics show that fewer meals are prepared, cooked, and eaten at home. In short, customers and consumers require little or no knowledge about foods, their preparation, storage, or safe handling or even, except in a general way, nutrition. Indeed, Toops (2000), an editor of *Food Processing*, has stated in reference to the U.S. situation, "… by 2001, most women aged 25-40 won't know how to use a conventional oven."

What are the obvious trends? More meals will be eaten out; more home meal replacements will be developed; more semi-prepared meals or prepared meals will be purchased and assembled and reheated at home. All in all, a victory for the food technologists and the food industry. But this does not translate into better health, better knowledge of food preparation, or safer handling of food. Indeed, malnutrition is being seen in many developed countries because a growing percentage of the population is becoming overweight and subject to the diseases that the overweight condition brings.

The growing acceptance – some have called it a love affair — of technology has now seen the emergence of the technocrat as a part of a ruling elite who base all decisions on appeals to their god, Science. If it is good for science it is therefore, no must be, good for everybody. But

2326/ch01/frame Page 37 Thursday, December 14, 2000 2:55 PM

If the past millennium has shown anything it is that nothing is constant but change itself. The changes that have occurred and problems that they have begotten for the food microcosm will be discussed in the following chapters. Maugham's words which close this chapter should be kept in mind because what has gone before will surely illuminate what is to come.

"That men do not learn very much from history is the most important of all the lessons that history has to teach."

Aldous Huxley

"Tradition is a guide and not a jailer."

W. Somerset Maugham

Chapter 2

Customers, Consumers, and Consumerism

The consumer isn't an idiot; she is your wife.

David Ogilvy, co-founder
Ogilvy & Mather, advertising agency

An Introduction to Change

The market economies of many countries have undergone a profound conceptual and actual change in the years preceding the new millennium. For some nations it has been sudden and abrupt change, for others a more gradual one. Some nations, the so-called emerging nations, were attempting to build their industries up to productivity levels on a par with the developed nations. Their attempts to develop their resource bases have not always been successful.

Elsewhere, the totalitarian nations are having to convert their economies from ones directed and driven by the needs and demands of the state to economies in which the demands of consumers must be attended to. The reasons for this conversion and for the degree of upheaval that was caused were varied and differed with each nation. However, since

made these changes felt by all industrialized nations. Hence, they have had a significant worldwide impact.

Both China and Russia are seeing this changeover to a consumer-oriented economy. The disruptions to what had been seemingly orderly state-driven economies are accompanied by major problems as their industries rethink their strategies, literally having to rethink what businesses they are in, review their capabilities, and retool their plants accordingly. It is a disruptive, even revolutionary, change; it is not an evolutionary one.

The whole focus of marketing must veer from satisfying the needs of the state to meeting the demands of its consumers and understanding the needs of consumers in other countries so that export markets can be developed. Innovative marketing programs are needed that put a priority on obtaining both a knowledge of the needs of the consumers and an understanding of what motivates consumers so that successful products can be developed. Then creative advertising and promotional programmes can be put in place to attract customers and ultimately to induce these customers to become regular consumers.

These changes beget others in ancillary areas. An entire infrastructure is required to serve the new consumers. It requires suitable retailing outlets and a well-functioning warehousing and distribution system to get goods to consumer outlets on time and in good condition (which introduces the need for a well-developed transportation system). It also involves such support systems as effective advertising in areas where the various elements of the media may not be available to the general public or where effective use of the media must be learned in order to reach the desired target audiences.

The above is, of course, an oversimplification of the activities required in the changeover from a state-run economy. The same problems exist for emerging nations. They too have to develop their marketing skills and their agricultural and manufacturing production capacities. The developed nations while they have more mature agricultural and manufacturing capabilities still need to maintain high levels of marketing and consumer research to remain competitive and keep their market share.

Some Problems for All

For food products the changeover to a consumer-focused system brings

- First, the nutritional needs and health of the populace must be determined before suitable agricultural policies and dietary guidelines can be put in place. Such policies will encourage, or even dictate, the adequate production of those crops and animals necessary for these nutritional needs and also necessary to provide surpluses for storage in times of crisis and for export to earn money.
- Then, governments need to develop sound agricultural policies. This development will involve considerations of trade policies, health and welfare of the population, and agricultural support policies for their farmers. These take time to develop properly.
- Food production requires a growing season. Crops must be sown and nurtured until they reach maturity; animals for meat production must be housed and fed until they reach market size. Then crops and animals must be harvested and stored until processed. The change from state-run cooperatives to privately owned farms is a difficult one that is not easily done rapidly.
- Because food is a perishable commodity, its preservation in some manner during and post-harvest is necessary: a technology-based infrastructure able to undertake this task is required. Where attention has never been paid to consumer-driven demand but only paid to state-dictated demands, this technology is hard to put in place quickly.
- Raw materials from farms and seas must be manufactured into desirable products, distributed, and made available to the populace at a price they are willing to pay. This requires that an infrastructure of manufacturing, distribution, and retailing capable of doing it, be in place.
- Finally, legislation must be in place to regulate every facet from the primary producer to the home to ensure the safety and purity of the food, fairness in trade practices, and truth in advertising and promotional materials.

Each step in the chain to provide consumers with products that satisfy their nutritional, emotional, psychological, and economic needs introduces complexity. As those aspects of nations' economies that are related to food and agriculture become more consumer focused, competition within nations for the attention of consumers will become more intense. Food manufacturers will vie with one another to get customers to buy their

products will be purchased abroad. Export sales of value-added food products are an important national asset.

Getting to Know the Protagonists

At the end of the previous chapter, some shifts of power were hinted at. A trend to the empowerment of the consumer was occurring. The consumer, and to a large measure the customer (the customer and the consumer must be considered distinct entities, *vide infra*), were becoming increasingly powerful forces in the food chain in the food microcosm. For example, in the food manufacturing community, the adoption of the ISO 9000 series of quality standards (International Standards Organization) made the buyer (a customer) a very important figure who could make very specific demands on the manufacturer.

This shift, from one of 'buyer beware' to one of manufacturer (and seller) 'take care,' marks a very important event which has many implications. Buyers are making demands upon their suppliers (i.e., upon manufacturers, distributors, and food retailers) and all are giving heed to these demands.

The implications of this empowerment of the customer (i.e., the buyer) must be clearly understood by both primary food producers and food manufacturers of added-value finished food products. Understanding the habits and needs of buyers becomes very important to fashion products they are attracted to. The demands of buyers will be passed down the line to the primary producers to supply that which the manufacturers want to those who sell to buyers.

Whether this shift in power is permanent remains to be seen in the early years of the third millennium. Already there are signs that further shifting is occurring.

A Clarification

Some explanation of the terms that will be used in the following discussions and chapters is necessary. A convention has been adopted to assist in the understanding of what is happening with the entity that is broadly described as consumers. The convention permits the discernment of the protagonists within the food microcosm.

they are in harmony with one another and sometimes they are not. It is not a fiction, therefore, to consider them as separate entities.

"When I use a word," Humpty Dumpty said, in rather a scornful tone, "it means what I choose it to mean — neither more nor less."

From *Through the Looking Glass and What Alice Found There*
By Lewis Carroll

It is important for all the protagonists in the food chain from the primary producer onward to be aware of the distinctions and similarities. Awareness of these facets allows a clearer understanding of how the balance of power between customer, consumer, retailer, manufacturer, and primary producer has changed agronomy, food processing, marketing, and selling as well as the various marketplaces in which these activities take place. This knowledge can then be a tool for what might change in the future.

The Customer

The term *customer* will be used to describe the purchaser of food, i.e., the purchaser of an ingredient, of raw produce or of primal cuts of meats, as well as of partially finished product or of a finished product. The customer purchasing food can have several very different identities. It is important to recognize these different identities in order to use the proper marketing procedures to find, to reach, and to communicate with the customer. Thus there are the following definitions:

- A food manufacturing company: One company buys a product from a second company. Company B's product will be used as an ingredient or component in company A's product often with no physical transformation. For example, the Néstle Food Company buys its pretzels from Recot, Inc. for making a chocolate or white fudge-covered pretzel (Pszczola, 1998).
- A chef, a professional cook, or a dietitian in some institution: All will purchase or recommend for purchase or prepare standards for the purchase of a variety of produce, ingredients, or semi-finished food items such as soup bases or concentrates to prepare a menu

44 ■ Food, Consumers, and the Food Industry

- An individual, for example, a mother, purchases food to feed her family. Here ingredients, finished prepared foods, or even take-out meals may be purchased;
- Pet owners, veterinarians, kennel owners: All purchase pet food but each has a slightly different reason or motive for their purchases.
- Medical doctors or professional nutritionists: They provide, for a fee, specially prepared dietary foods (which they have purchased from specialty food manufacturers) for those with unique nutritional problems.

Each of the above purchase food or they are responsible for the purchase of food to be prepared for others. They are customers.

In each instance above, as customers:

- They have the *choice* of what to buy. Whether they be teenagers purchasing finger food, housewives buying the family's weekly groceries, or purchasing agents of large food plants, they control decision-making respecting what is purchased and the amount of money that is to be spent.
- They or their agents are the persons influenced by the forces of the marketplace such as sample tastings, demonstrations at trade shows, promotional materials, and advertising. They receive visits, gifts, and other blandishments from sales representatives.
- They are the targets for sellers of produce, ingredients, or added-value food. Retailers and food manufacturers alike target them with inducements to buy. The customer is the center of their interest.

The power to decide what to buy lies then with the customer. The customer has, in addition, the power to decide what others will eat. Even when manufacturers direct their products at children, the customer, the parent in this instance, has the power to say yea or nay to the wants of the children. To describe an individual or a manufacturing entity as the customer is to admit that that individual or entity controls the purchases of food and is subject to all the blandishments of the retailers, sales personnel, and food manufacturers in the marketplaces where the food is sold.

The customer is the gatekeeper. This is a concept first developed by Dr. Kurt Lewin (reported in Gibson, 1981). This term was originally used

food for use or consumption by others or that is processed by another entity and given added value.

Many marketing professionals, to their peril, forget this concept of the customer as a gatekeeper. This concept has an impact on market research and hence on new product development, on the development of promotional materials and advertising, and on selling.

The Consumers

Simply put, consumers consume; that is, they eat, or in the broader context, use up that which was purchased or prepared for them.

Customers use products in the preparation of other food products. Food manufacturers make ingredients that another company or an individual uses in the manufacture of products that are sold to other customers down the line or fed to consumers. The chain can be quite complex; a manufacturer makes an ingredient that is sold to another manufacturer for use in its finished ingredient. This product is in its turn used by yet another manufacturer for another product until ultimately one manufacturer produces a product destined for a consumer to eat. Each stage adds a refinement, an added value, to the original manufacturer's product. They suggest a series such as manufacturer, customer, customer…customer, consumer.

How then do consumers wield their power in the food microcosm? First, they influence customers, i.e., those who do the buying, for example:

- Children can have great influence over their parents about what breakfast cereals, cookies, or snacks are purchased in the household. Parents are also influenced in their other food purchases by their children's personal likes and dislikes or their allergic reactions to foods.
- Prison riots over the quality of food served to inmates have caused prison officials to rethink their menus respecting the dishes, the variety, their quality, and quantity.
- The loss of passengers who depart one airline carrier because of poor food to go with another which serves better food will have its repercussions in ticket sales. The senior management of the airliner losing passengers over poor food might review its policy regarding its in-flight menus.

■ Purchasing agents favor those suppliers who deliver on time
   with consistently good uniform products meeting all the desired
   standards.

Second, by their very nature, consumers do form recognizable marketing niches. These are exploitable. That is, they provide selling opportunities. Food manufacturers, through their market research can identify
them and prepare new products that satisfy the specific needs of that
niche. Then, the marketing departments of these manufacturers create
promotions and advertising to make the targeted consumers aware that
these products fill that need. Then they promote so as to attract the
segment of the population they have identified.

The obstacle to this is, of course, the customer. The consumer must
go through the barrier posed by the customer: There is the conflict
between the consumer's hedonistic demand of "I want" and the customer's
practical barrier of "I need," or "I can afford" or "Ingredient costs must
be kept within this cost figure" or "This product fits our requirements
but not that one."

Manufacturers and sellers alike must placate the customer yet please
the consumer.

The Hybrid: The Customer/Consumer

There is an obvious overlapping gray area in which customers are also
consumers. A parent will, in all likelihood, be both the customer for the
family and a consumer while other family members are only consumers.
Customers in a fast food restaurant are also very likely to be consumers
of the menu items that they buy. Purchasers of finger foods are also likely
to be the consumers of that food as they stroll down the street.

Despite the distinction made between the customer and the consumer
in the previous sections, the hybrid, the customer/consumer, someone
who buys and eats, inevitably arises. Thus, the three entities, customer,
consumer, and customer/consumer are different in a food-marketing sense
yet they are subtly interwoven.

Marketing professionals cannot treat customers, consumers, and the
hybrid as one. Certainly they must recognize the similarities but they must
also be aware of the stark differences among the trio if they want their
marketing strategies for the new marketplaces in the new millennium to

base their new products. The advertising and promotional programs they conceive must also target the correct audience. The simplest example would be food products targeted to children. These must attract and satisfy the child who is the consumer. They must also impress the parent (or at least not cause distress) with their nutrition and wholesomeness: The parent is the customer. Advertising and promotional materials attract, but in very different ways, the parent and the child. This becomes very difficult where advertising directed to children is not permitted.

Manufacturers of industrial food ingredients meant for use only in the trade must target their audiences very differently. They have two targets: the industrial user and the consumer. For industrial users, promotional activities are very different. For the consumer, they must demonstrate the benefits of the ingredient in the finished product so that a demand is created that induces the industrial user to adopt the ingredient. Manufacturers of snacks and finger foods target the hybrid customer/consumer.

Thus, there are real differences between customers, consumers, and customer/consumers which those in the food microcosm must recognize and accommodate.

The Impact of Change During the Past 1000 Years

The profound changes in the political, technical, and scientific arenas in the past 1000 years in their turn have caused changes particularly in the food arena. For one thing, there has been a profound growth in the power of the food consumer in this same time period. There is now, for example, gray power to describe the influence and consuming power of senior citizens who have money and leisure; and teen power, to describe the consuming ability of teenagers and the capacity they have for causing manufacturers to satisfy their needs. There are also the baby boomers, a name coined to describe people born into an era of affluence with very specific social and cultural habits and power as consumers. And there are such classic acronyms as yuppies (young urban professionals) and dinks (dual income no kids) to describe the buying power of the very materialistic young professionals of the 1980s.

Neither the consumer nor customer was previously a strong force in the marketplace. This empowerment is a very recent phenomenon and it has yet to prove its permanence. To say that this is the age of the consumer

A Power Shift

From the start of the past millennium and for the greater part of it, almost to the 1900s, the balance between primary producer, manufacturer, retailer, and the customer/consumer was fairly stable. Figure 2.1 depicts figuratively what the situation was for these first several centuries. It was a largely agricultural-based economy. The primary producer was at the top of the pyramid, supreme in importance and king-of-the-hill. Often the primary producer acted in part as both the manufacturer and the retailer of food products. In the early centuries of the last millennium food manufacturing was done usually at the farm level. It was rudimentary processing, primarily pickling (curing), salting, smoking, and drying. In many ways the consumers, as well as the primary producers, were also manufacturers in a sense and "did down" by preserving the family's (or the manor's) food by salting (corning), drying, fermenting, cheese making, and so on. Brewing was often done at home. Baking was a retailing operation.

The customer/consumer was literally at the bottom of the pyramid during the greater part of the past millennium. There were no efforts to analyze their buying habits, no surveys or opinion polls to determine their preferences for particular products. Indeed, toward the end of this period, manufacturers and retailers alike, in many instances, considered customers to be fair game to be bilked. The health and safety of customer and consumer were not given a thought.

The situation at the end of the millennium and in the early years of the new one, Figure 2.2, is almost topsy-turvy to the earlier years depicted in Figure 2.1. The customer/consumer is dominant. It must be clearly understood that the customer is still a barrier (the gatekeeper) to consumers. While customer/consumers may be a very fragmented group they are emerging as a dominant force in the chain. Retailers (sellers) at every level try to entice customers with promotions, sales, convenience, and most importantly, service. This service may for industrial customers be prompt deliveries of uniform quality products and for the individuals service may be home delivery of groceries, clean and bright stores, and cheerful attendants. Retailers study their customers to determine how to attract and keep their customer base. In short, the customer has been empowered.

The customer/consumer entity has power of choice, in other words, the power to choose any food from an array of foods. This being so, the drive by sellers, then, is to understand the psychology underlying food

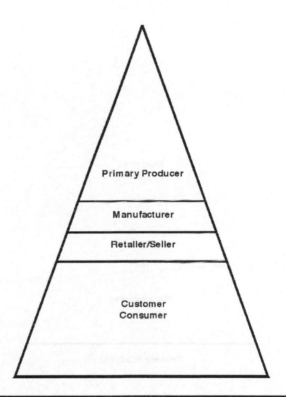

Figure 2.1   Relationships among the Protagonists of the Good Microcosm during the Major Portion of the Second Millennium

- The individual's personal concepts of food safety, i.e., what is safe according to his/her state of health, food sensitivity, and food aversions (not necessarily allergic responses).
- The composition of the diet especially respecting protein, fat, and carbohydrate composition.
- Family religious and ethnic tradition can influence choice. Tradition can have both a religious and ethnic impact on food choice.
- Social factors or status; food can be a status symbol. For adolescents especially, peer pressure can affect the choice of what is acceptable to eat.
- For children and teenagers, role models (sports athletes, older siblings who are looked up to and who eat, for example, their vegetables) can sway food choice. Reward systems for eating

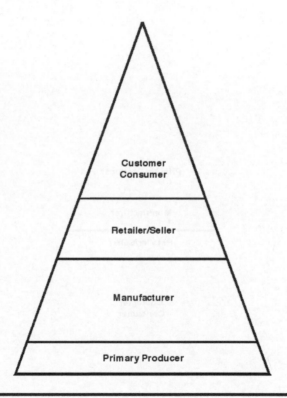

Figure 2.2   Changed Relationships among the Protagonists of the Food Microcosm During the last Years of the Second Millennium; the Dominance of the Customer/Consumer Begins the Third Millennium

Retailers and food manufacturers alike are very keen to unlock the keys to what drives food choice.

Retailers, especially the large chains, have gained enormously in importance (Figures 2.1 and 2.2) and are quite able to dictate their product costs to primary producers and buying price and delivery requirements to food manufacturers (see Chapter 4). This power permits them to control their retailing costs. Retailers can dictate advertising and promotional requirements to their suppliers for the various promotions they will undertake.

Manufacturers are gaining in influence (Figures 2.1 and 2.2) but are still subservient to retailers, especially the large chains. Manufacturers can simply pass these pressures on to their primary producers. Multinational

The primary producer or gatherer has gone literally and figuratively to the bottom of the pyramid (Figures 2.1 and 2.2). They cannot always keep up with the demands of either food manufacturers or of retailers. Large retailers, for example, can put pressure on poultry producers and processors for fresh chickens and chicken parts in barbecue season telling the producer what price the product will be delivered at and when for promotions.

Primary producers are at the mercy of weather conditions which can greatly affect the availability of produce. Unfortunately, promotions and sales march to a different beat. They are timed to go with precision and supplies had better be in place to meet demands.

It is not only the retailer who can dominate the primary producer. The manufacturer's power over the primary producer was demonstrated forcefully to me at a conference I attended sponsored by the Guelph Food Technology Center and the Ontario Egg Producers Marketing Board entitled *Expanding the Egg Horizon* held February 17, 1997. One speaker, Michaelides (1997) stated plainly that the costs of production of egg-containing products in the U.S. were 1/3 the cost in Canada. The main reason for this greater cost was largely due to the greater cost of eggs and egg ingredients. (Egg supply is closely controlled in Canada through marketing boards.) Research was actively being conducted to reduce production costs in Canada by finding substitutes for the egg ingredients or by eliminating eggs from their products. Unless the costs came down, his company was prepared to import finished product from the U.S. into Canada. This is a good example of a manufacturer of egg-containing food products demonstrating its clout to a supplier/retailer.

The Cult of the Consumer: Consumerism

This striving on the part of retailers and manufacturers to gratify the needs and desires of customers and consumers has had some interesting results. One result has been the growth of consumerism. Nowhere is this consumerism being more strongly felt than in the microcosm that is food.

First, customers and organizations representing them are realizing their united strength, and are gaining awareness that they can make demands and their voices can be heard. They have power. In effect, consumers and customers are telling food manufacturers and retailers alike what they want. If they do not get what they want, they will find it elsewhere from

52 ■ Food, Consumers, and the Food Industry

to activists to demonstrators to alarmists to what? Terrorists? One can remember violent demonstrations or activities such as the poisoning of grapes to protest unfair migrant working conditions, poisoned turkeys and spray-painted meat products acts committed by animal rights groups, displays of irradiated product being overturned, or fields of genetically altered produce being destroyed by masked activists.

Consumer activists have compelled government agencies to audit their own activities to improve their services to consumers (Bell, 1999). An audit of meat inspection in Canada showed falsified records, mistreated animals, and inspectors ignorant of their own regulations — the auditor requested 40 corrective action requests.

Webster's New World Dictionary, Third College Edition provides three meanings for consumerism:

"1, the practice and policies of protecting the consumer by publicizing defective and unsafe products, misleading business practices, etc.
2, the consumption of goods and services
3, a theory that a continual increase in the consumption of goods is sound commercially."

Most people when they hear or read the word 'consumerism' think primarily of the first meaning and then mentally visualize some association or outspoken zealot decrying some product or activity.

The darker side of consumerism is a militancy that sees an anti-trust attitude building toward activities in the food microcosm from "meddling unnaturally with food" to misrepresentation, to false claims, to adulteration, to cruelty to animals. It gives every evidence of conflict in the future.

A recent newspaper describes a report that appeared in the *Washington Post* about a conflict between Asian Americans and animal rights activists over the sale of live food. Animal rights activists allege that live animals are dismembered prior to killing to demonstrate freshness (Stern, 1998).

Again, in the December 12, 1998 issue of *New Scientist* (Anon., 1998a) a brief article described Britain's desire to cull 20,000 badgers to reduce the possibility of them infecting cows with bovine tuberculosis. Consumer advocates and dairy farmers who want safe(r) food applauded the government's action. Needless to say, animal rights activists and environmentalists were upset by the planned action.

All in the food microcosm condemn the presence of salmonella in

There is a fear that irradiation will be used to hide, so to speak, poor handling practices.

There is a drive by some activists to have food products that have been irradiated labelled as such. Several problems arise should this need to label irradiated foods be endorsed:

- There will be a cost to inspect and analyze product to determine if it has been irradiated
- Inspection, sampling and analyses will delay delivery to consumers. Products awaiting test results will need to be stored
- Current methodologies for detecting whether products have been irradiated are not equally sensitive in detecting low levels of irradiation. There are no routine standard methodologies

These few examples of the power of consumerism and consumer group activities demonstrate the sorts of conflict that can arise. These will not soon abate; indeed, there is every indication that they will intensify.

There is a final result of consumerism. The second and third meanings of consumerism, namely, consumption of goods and services and any continued increase in goods and services are good — good for the economy. Therefore, there is a great rush to understand the customer/consumer entity for profit. When such knowledge and understanding becomes so comprehensive as to become manipulative of the customer/consumer, some backlash from the customer/consumer can be expected.

Techniques for Understanding Customers and Consumers

Much information about customers and consumers can be obtained by simply observing them. There are two kinds of observations of customers and consumers:

1. Measurements. A variety of measurements can be taken. People can be counted; their family incomes can be recorded — that is, government census data are available to describe the composition of various communities throughout a country. Per capita consumption figures from government sources give crude estimates of what

2. Real observations. Consumers can be observed directly by unob-
trusively watching them or by using hidden cameras to catch their
shopping habits and reactions to tasting foods. Their paths through
the grocery store are tracked. Areas where they are attracted, or
where they are attracted *and* buy or sections that they bypass
entirely or visit less frequently are carefully noted. Their reactions
in groups to product concepts can be evaluated. Their values and
attitudes are assessed through surveys.

The two are crudely separable into demographic data and psychographic data. Measurements, demographic data, pertain to people and
describe the physical attributes of populations. Psychographic data attempt
to analyze the habits, values, motivating factors, and attitudes of specific
populations. Combined, interpretation of both sets of data paints a more
complete picture of the dynamics of customer/consumers.

Consumer research has progressed rapidly after a slow, even disinter-
ested, beginning in the late 1900s. Disappearance figures of foodstuffs,
case movements of selected food categories, and dollar volume of store
purchases were the main tools used to determine the food habits of
consumers and uncover any trends in their buying habits.

As competition for customers grew and competition between food
manufacturers grew, the need for better and more accurate information
became more critical. Mass marketing was superceded by niche marketing.
This required more intimate knowledge of the habits of smaller groups
of customers and better means of getting the data. Technology became
more sophisticated and the techniques much more refined.

Customers have been both observed and questioned to assess their
shopping habits and also to define their needs, desires, likes, and dislikes.
Observations of their buying habits were vital to determine what attracted
customers as they shopped. Demographers and psychographers analyzed
the mountains of data that became available and arrived at some very
precise descriptions of consumer types and habits.

On the threshold of the third millennium, getting to understand what
value systems customers and consumers use to make their choices; to
know where, why, and how they shopped; and to develop the best means
to communicate with them are essential tasks. Each methodology to do
this brings its own degree of success if carried out properly and if the
results are interpreted rationally. The limitations of each technique must

To understand and anticipate some of the problems that are arising it is necessary to recognize the techniques for understanding the customer/consumer. The knowledge that is being gained is beginning to alarm some. Many groups are concerned about how personal information that can be gained is being used. They fear that privacy and human rights are being violated. They also fear manipulation by manufacturers and retailers alike.

Measurements for Consumer Research

What customer/consumer information is required? What information is needed will vary with who wants the information and for what purposes they want it. Food manufacturers want to know such information as:

- Who is the customer and for whom do they buy? It is extremely important for manufacturers to know who their customers are. What do they buy? Where (in what type of outlet in what geographic location) do they buy the company's product? When (a seasonal effect?) do they buy it?
- How and where do they use the products they buy? Do they buy in one nearby location and use it in another distant location?
- Do the products they buy suggest other products that might be even more desired? That is, does they way they use the product suggest new ways to promote new and/or added-value products? Companies need to replace existing and aging products with new products; and, as well, companies want to know
- How strong are their brand franchises? What do their customers think of the brand? How do they evaluate the brand? This knowledge allows manufacturers to determine how widely their range of products can be expanded.

The desire for more customers and for a greater market share for their branded products — the goals of all marketing departments — drives a need to investigate more and more what motivates customers in food marketplaces to make purchases as they do.

Retailers have slightly different objectives. They want to know:

- Are there any peculiarities in the habits of shoppers that can be

- How can stores be redesigned to make shopping a pleasure and lessen irritation to customers yet provide greater exposure to products?; and
- Where is there the greatest density of potential customers for determining the most likely places to site a fast food outlet, or a coffee house, or a restaurant, or a bistro, pub, bar, or a shopping mall.

What are the tools to accomplish these goals? Gathering information about the customer/consumer entity need not be a cumbersome and expensive operation. Simple techniques can provide good information if they are used wisely.

Information from the Sales Personnel

A company's sales personnel as they make their rounds in the particular marketplaces that they service (see Chapter 3) see what sells or what customers take an interest in. This simple technique of reporting back what sells and what interests customers is effective in shopping malls, mom-and-pop stores, and at trade fairs. It borders on being psychographic data. A colleague, a senior vice president of marketing, visited different types of food stores and outdoor markets whenever and wherever he traveled. He talked to clerks stacking shelves, to store managers, to customers wherever he could be understood, and he observed competing products, their facings (he would pace these off to get crude measurements), and promotional materials used by competitors.

Visits by sales personnel reveal what the competition is doing, provide an opportunity to meet customers and to discuss their needs. Sales personnel can also network with retailers and so bring back ideas to their companies that help the retailer and coincidentally improve their own sales.

Telemarketing

A 1-800 number provides food companies with an opportunity to gather information directly from customers and consumers. By dialing a 1-800 number, a customer/consumer can talk with a company representative and unwittingly reveal a great deal of personal information useful to consumer and market researchers. A skilled customer service representa-

When food companies subscribe to a 1-800 service, their service providers give them a list of the telephone numbers of their callers. These numbers can be used with reverse directories to identify the caller and where they are calling from. Very often this is his/her place of residence.

Then with comparatively easily available demographic data that describe the district where the caller lives, the food company can identify the economic status of that individual, i.e., people of like economic status and preferences generally congregate in the same areas. By correlating all the available data from all callers, a massive amount of information can be gleaned. Food manufacturers can then develop products that fit the demographic niche or they can target a better product mix into a particular geographic area. Instead of a shotgun approach to customers and consumers in a particular area, a more targeted approach can be attempted as the database is developed.

This becomes an even more exciting tool when the above is fully automated. That is, when the 1-800 number is dialed, the representative has the destination of the caller immediately identified. Software correlates this identification with a knowledge-base capability founded on demographic data which map geographic areas by income, by ethnicity, by age, by land and house values, etc. The customer service person now has generic information with which to converse with the caller and more accurately direct him or her to other products or services the food company offers. This practice is referred to as cross-selling or up-selling.

Computer telephone integration in the hands of a skilled, trained representative processing incoming calls provides a wealth of information such as:

- A catalogue and categorization of customer/consumer complaints about products or promotions in real time
- A collation of ideas and suggestions from consumers about how they use the company's products or what they would like the company to offer as products
- An opportunity to advise consumers about recipes using the company's products, to give nutritional information about products, to provide general company information, or to answer any questions consumers have about using the company's products
- Improvement of products or promotions based on an analysis of problems consumers report to the customer service representative

Cooperative Data Sharing

Retailers possess a great advantage for understanding their customers; they can catalogue everything the customer purchases in the sales receipt. The store's copy of the sales receipt contains a wealth of information that can assist an understanding of customer habits. This simple piece of paper provides data on:

- How much money the average customer spends per shopping visit
- When that visit was made (supplying both the time of day and the day of the week) and
- What combinations of food and non-food items were purchased during that visit. The food items can be further broken down into what combinations of fresh produce, frozen foods, dry goods, pet foods, meat and fish, etc. were purchased together.

Much can be learned about the customers in any community when data about the immediate area in which the store is located are coupled with demographic information that can be found in census data and which is readily available; for example,

- The average disposable income in the region where the store is located. Is the region a professional or a blue-collar community? More up-scale products would be displayed in a wealthier community to complement the existing range of products.
- The average age of the residents of the community. That is, is the community a young community or is it a more mature community? Young communities can be expected to have young children. Promotional activities and products would be designed to cater to a younger and more active crowd.
- The ethnic make-up of an area. Products reflecting the diversity of the community's ethnic profile would be expected to predominate on store shelves.

The use of credit or debit cards by customers allows even more useful data to be ferreted out. These receipts identify the buyer with the purchased items. This data combined with general census information can provide the retailer with

A Summary

The above demonstrates how much information can be obtained quite legitimately about customers and consumers and how other data from other sources can be used to develop a profile of their lifestyles. Information is provided unwittingly by the customer and can be used by marketers to identify their customers' economic status, to understand their customers' buying habits respecting what and when they shop, or to learn what combinations of food and non-food items are purchased so that combination promotions can be designed to target their customers more directly.

Obviously food manufacturers are quite willing to purchase such information from their retailers for the marketing advantages it provides them. The concern arises — and it is a growing concern — that such information can be considered an invasion of privacy.

Getting into the Mind of the Customer/Consumer

These techniques are more sophisticated and require skilled professional personnel to both carry them out and to interpret them. Few retailers or food manufacturers have such expert staff. Consequently, outside help must be called in.

Many market and consumer research companies collect and process data which they obtain from a variety of research techniques as well as conduct surveys on their clients' behalf. These studies include assessing groups of consumers for their reactions and attitudes toward real products, product concepts, and advertising and promotions. Food retailers and manufacturers alike subscribe to such data-collecting services to gain much descriptive customer/consumer intelligence.

Focus Groups

Both quantitative and qualitative data about customer/consumers are desired by food manufacturers and retailers. For qualitative data a company may want to conduct a focus group. In these, the reactions of a group of people to a product or to a product concept which the company wants to refine prior to doing intensive and expensive development work can be studied.

60 ■ Food, Consumers, and the Food Industry

For products targeted at teens and adolescents, obviously teenagers and adolescents will be selected; products for females will have only female participants, etc. Occasionally a sampling of consumers selected at random is used.

Research companies that conduct focus groups will have their own lists of potential interviewees filed according to the specific consumer characteristics the research company requires. The company then randomly selects from the list of candidates with the desired attributes. Thus a group of seniors, for example, is randomly selected from the research company's list of candidates who fit the seniors category for age. Since candidates who take part in the focus groups are usually paid for their trouble, consumer research companies should make some effort to avoid so-called professional interviewees.

A moderator trained in group dynamics conducts the session, demonstrating the product or product concept and probing the group for reactions and comments to this material. The sessions are often videotaped for review and analysis of the proceedings later. This analysis can be a very thorough undertaking using professionals to analyze body language, mannerisms, and language to delve into attitudes about the subject matter.

Often, the client whose product is under review and the research company's personnel observe the reactions of the interviewees through one-way glass. The client and the professional analysts can discuss their observations together as these are occurring. Rarely are any conclusions based on the results of one focus session. Several focus groups, therefore, will be conducted for a client.

The results of any focus group study do not have any quantitative value. The data have only qualitative value. The interviewees do not in any respect represent a cross section of the general population nor are they representative of the unique (targeted) population under review. The samples are too small and interpretation of data while done by professionals is still subjective in nature and is unscientific. Consequently, there is no improvement in the information that is generated by conducting more than half a dozen focus groups on any one subject. Their value is primarily for the generation of ideas they may spark or for an evaluation of a company's preliminary concepts for new products that may occur. They provide only a very general direction for further work.

Focus groups are, however, not an invasion of privacy.

toward foods and food purchasing. They are used to predict trends. Or the data so collected from surveys can be used to determine associations of events. That is, they uncover things such as:

if "x", "y" and "z" happen, then "w" will occur with a probability "p".

While both results are useful to food marketers the second is rather more exciting. For example, knowing that a customer purchases certain items together, marketers can position their products, for example, the "w" above, to be close to the x, y, and z, or co-advertise with these products, or piggy-back with them.

Surveys are, however, very tricky to carry out properly. How they are prepared and executed can greatly influence the answers of the inter-viewees. The interviewer, consciously or unconsciously through manner-isms, clothing, body language, or even the wording and order of the questions asked, greatly influences the interviewee's attitude if the survey is a personal encounter. Timing of telephone interviews can bring negative attitudes from interviewees if they are intrusive.

Governments periodically undertake a census. The result is a descrip-tion of a population according to certain characteristics (i.e., the questions asked) at a given time. They are usually carried out at 10-year intervals. Thus population trends can be traced at 10-year intervals for as far back as the data are available. The data are readily computer-integrated to the telemarketing systems described above.

Broken down into geographic regions these data describe population changes in specific areas. Predictions of possible events for the next decade in those narrower areas can then be made. Data from a census survey such as ethnic heritage, age of respondents, individual and household income, education, family size, sex, hours worked per week, possessions (e.g., car, TV, etc.), house payments, rent, amount spent on food per household, provide valuable information for both food retailers and food manufacturers.

The data from such surveys are, and always will be, historical. The data were gathered "back then" and can only be used with care to predict what will be tomorrow. For example, to know from census data that the average number of children per family rose from 1.2 to 2.8 in a decade will clearly suggest that there may be a need for a school building program to handle the greater numbers of children, which may, in its turn, suggest a need

- Local government rezones a residential area to a commercial area and forces a migration of families out of the region.
- Technological change impacts the region such that local industries are forced to close or to relocate to new facilities taking families with them.

Caution must always be used in the interpretation of census data.

Governments provide other data which can be used profitably, for example, figures on food production and manufacturing volumes, on food imports and exports and so on. Other surveys are conducted on any number of topics by governments and by private firms. For example, many private firms provide information on the movement of case goods; that is, they provide data on the case volume movement of particular product categories within the distribution system on the assumption that this activity correlates with customer purchase habits — an assumption that can be misleading in predicting a trend.

Issues Confounding Knowledge of the Customer/Consumer

The customer/consumer is a complex entity whose understanding requires a juggling of many different observations and subjective interpretations of these observations. One enters into an area of soft science, a science not based on rigid principles. Successful retailing entrepreneurs with a 'gut feel for selling' are often just as successful as large companies with an extensive market research organization but, they, although successful, may not always know why. Codifying the *modus operandi* of successful retailers or food manufacturers may not provide general procedures suitable for the vast majority of marketplaces.

Studies on consumers and their habits and behaviour as they are actually shopping may offer more knowledge on what annoys them or what pleases them in the facilities where they shop. Such information is useful perhaps in the design of retailing outlets but sheds little light on the customer/consumer. Knowing that people turn away from a counter if touched or bumped on their backsides does little more than suggest that the aisles in stores should be wider.

recognizably similar" (Moroney, 1953). Neither consumers nor customers represent a homogeneous group, nor are they recognizably similar. They are composed of children who can be divided into babies, toddlers, tweenies, and teenagers; adults who can also be broken down by age groups, by social and ethnic groupings, and by activity (sedentary, laborer, sports activists, etc.). They represent the wealthy, the financially comfortable, and those on tight budgets. There are the health conscious, those on specialized diets because of disease conditions, and those consigned for religious or philosophical convictions to restricted foods. Consumers and customers must be broken down into more and more categories before any semblance of homogeneity begins to appear.

If the cult of the average is followed assiduously, knowledge of either the customer or the consumer is bound to fall abysmally far from its goal of providing an understanding of either the customer or the consumer which can be capitalized on.

In Figure 2.3 there is depicted a typical bell-shaped curve for a hypothetical purchase. The vertical axis represents the frequency of purchase of some product; the horizontal axis is some descriptive characteristic of the consumer, for example, age. The average is a statistical concept, i.e., the average age of consumers is that point on the horizontal axis where the curve reaches its maximum vertical value. The number of units purchased by this (hypothetical) average-aged consumer is the number of units (height) on the vertical axis at which this curve peaked. Thus one speaks about the consumers who purchase y units are, on average, x years old.

A lot of information is obscured by such a statement. Averages obscure the variability of the data from the population that is being studied. How broad is the distribution? They overlook (unless the data are plotted) how the data may be skewed or distributed. For example, a hot sauce may be

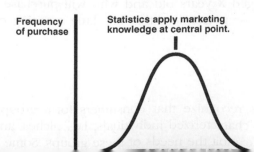

Frequency
of purchase

Statistics apply marketing
knowledge at central point.

used as both a condiment, i.e., applied to a prepared food dish, or used as a cooking ingredient in the preparation of a food dish. If a survey of how many bottles of a product are used in a given time period is conducted, an average may be misleading. Condiment users may use only one or two units per time interval; those respondents who use it regularly in cooking may indicate they use six or seven in the same time period. The average will be somewhere between two and seven bottles.

This manufacturer loses information about the customer/consumer. The bipolar nature of the data is lost; users are highly polarized. There are two distinct populations that must be served. Any action such as making a larger size unit or launching promotions ignoring the polarity of users may not have a beneficial return. The segmentation within their hot sauce market has been lost. One can use averages only when the set of figures one is analyzing constitutes a real family of data and not, as in this example, two distinct families of data from two very different user groups.

Thus, within any general set of data based on the responses of random encounters with customer/consumers, there will be subsets of data the interpretation of which will indicate very different users or market segments. Manufacturers will use very different product designs or promotional approaches to benefit fully from each segment.

Where and who then are the average consumers of, for example, high energy foods? They are not members of a homogeneous group; they are not recognizably similar. As to the where, the consumers of high energy foods may just as likely be found in a gymnasium, on a playing field, or in a convalescent home. Who they are defies description. They can be any age, any sex, any religion or philosophy, any social or economic class, and belong to any ethnic group.

The average customer or consumer is hypothetical. The term should only be used in such inane statements (see Figure 2.3) as "the average consumer who is aged X years old and who will purchase Y units of a particular product" — a statement that has little meaning or value in a marketing sense.

Niche Markets

Food manufacturers recognize that consumers form groupings of like-minded or similarly characterized individuals, i.e., niches, and they make special products to exploit the needs of these groups. Some categories of market niches with examples of types of foods for each are demonstrated in Table 2.1. None of the classifications are exclusive, that is, all food products, for example, snacks such as potato chips, can be, and are, consumed by all segments of the population but teenagers may be a dominant group (niche) to be targeted. High calcium foods or foods with calcium added to them (e.g., milk and orange juice) are consumed by the elderly and promoted by parents of teenage daughters (notoriously poor eaters) as preventive measures against calcium deficiencies. They may also be prescribed by nutritionists for older women with incipient osteoporosis. High protein or high energy foods may be prescribed for post-operative patients and may be used by athletes in training.

A Recapitulation

Knowing the customer and the consumer and understanding both are two very different processes. One can know someone or something in an abstract way. But a list of disconnected facts cannot bring understanding of that person or thing. To say that a person is tall, dark, slightly overweight and middle-aged tells something about that person but one still does not understand that person. Even if numbers are attached to the description that has been obtained; that is, this person is 5 foot 11 inches tall, weighs 210 pounds and is 48 years old, more data provided — one might recognize the individual — but little information on which to develop any understanding of who this person is, what his/her preferences are, or what he/she is likely to purchase. Measurements to which numbers can be attached do not bring greater understanding. The consumer who is only, demographically defined, that is, defined by numbers only, is a poorly defined entity.

Psychographic data are needed in conjunction with demographic data. This begins to describe the habits of people: How do they live? Where do they live? Where do they congregate? How do they spend their leisure time? What do they spend their disposable income on? What magazines and newspapers do they read? Now a picture of the customers and consumers begins to emerge and they become known individuals.

Perhaps too well-known. Many object to the invasive tools that marketing people can use legitimately to define the habits and characteristics of their target, the customer/consumer, for a surer hit. Are there real concerns about manipulation and an invasion of privacy?

Table 2.1 A General Classification of Marketing Niches where Customers and Consumers with Unique Food Needs May be Found

Classification	Marketing Niche
By age	• Seniors: by lifestyle and by nutritional requirements (active, limited capability, special nutritional needs based on infirmities)
	• Middle-aged: dietary foods (low fat, low calorie, high calcium)
	• Young Adults: by lifestyle and activity (foods for an active lifestyle, nutritious snack foods, calcium and iron fortification for women)
	• Teens (nutritious snack foods, finger foods, very active lifestyle, calcium requirements for girls)
	• Tweens (nutritious snack foods, finger foods)
	• Children (nutritious snack foods, finger foods)
	• Toddlers (introductory foods)
	• Infants (baby foods, etc.)
By income	• Upper quartile to lower quartile
	• Above or below poverty line
	• Specific salary ranges
	• Dual income or single income families
By dwelling or location	• Urban, suburban, or rural
	• Cities and towns by population size
	• Apartment dwellers or single unit homes
By physiological status	• Sex
	• Pregnant or lactating female
	• Physically handicapped: by nature of handicap
	• Convalescent (short or long term)
	• Allergy prone
	• Immunologically compromised
By ethnic background or cooking style	• Chinese, Italian, Mexican, Thai, Lebanese, Japanese, Indian, French, Cajun, etc.

Table 2.1 (Continued) A General Classification of Marketing Niches where Customers and Consumers with Unique Food Needs May be Found

Classification	Marketing Niche
	• Military feeding: portable foods designed for military maneuvers under intense activity and stress • Athletes: foods for short-term intense activity or long-duration activity (high protein, high calorie, isotonic beverages, nutritional supplements) • Leisure activity: hunting and fishing, camping, hiking, picnicking
By religion or philosophy	• Ethical foods: foods produced by companies with ethical principles respecting the treatment of animals (anti-factory farmed animals), the environment, human rights, employment practices (equal opportunity employment), workers' wages, etc. • Organically produced and processed foods • Vegetarian foods: several different niches are possible • Foods to satisfy various religious requirements of which Jewish and Muslim rites are the most dominant • Foods prepared according to philosophical traditions: Yin and Yang, Zen cookery

Chapter 3

Marketing and Markets

"Sell cheap and tell the truth."

Rose Blumkin

"Advertising may be described as the science of arresting the human intelligence long enough to get money from it."

Stephen Leacock

What's in a Name?

Most people know that somehow a food product gets from the food manufacturer to the customer and ultimately to the consumer. How this happened or what intervening steps took raw products from the farm or the sea and fashioned them into a desirable product that was needed, they are less certain about. They are unaware that from the fields and oceans, to the table:

- A food company through its consumer and market research uncovered perceived needs of customers and consumers which presented marketing opportunities, and the food manufacturers developed products to fill these needs (market niches)

■ Customers then had to be informed of the presence of this product in a manner to attract their attention. Certainly the attention of consumers had to be engaged in order that they could direct the customers' attention to what they, consumers, would like.

Several terms were introduced in the foregoing: marketing, market and niche market (or market niche), and marketplace. Also mentioned were references to consumer research and to the necessity of informing customers and consumers about the products that become available.

Most people, including one vice-president of marketing whose ignorance of the term was the cause of his removal, have only a vague idea of what marketing is. "Marketing" and its congeners "market" and "marketplace" are, through this ignorance, much maligned and misunderstood terms. When asked what marketing is, many would answer the question with another query: "Is it selling?" or, "Is it advertising?" A perusal of many books on marketing will show that many others are equally confused and restrict the word "marketing" to a very narrow interpretation, to wit, advertising and promotion.

If most food executives or technologists were asked to define marketing as an activity within their companies there would be diverse descriptions ranging from the activities of "sales and selling" to "markets and marketing," to "advertising and promotions." To be fair, although these are different activities there is a common thread linking them.

Most people use the words, sales'n'marketing or marketing'n'sales, as if they were one, spoken in one breath. Even marketing people themselves cannot explain clearly what distinguishes sales and marketing without resorting to merely providing a description of what they do in their company. Indeed, the executive, above, who was released could not clearly distinguish the two activities — selling was marketing in his mind. He was a great sales executive but not a competent marketing one.

The confusion is understandable. In many smaller food companies, the sales manager is also the marketing manager. Rarely is there any distinction made between the two functions within these smaller entities. Nevertheless, marketing is not selling; selling is not what the marketing department does.

Many people when they hear the word "market" mentally visualize a physical place, a mall or a supermarket, where vendors sell products and where customers can wander around inspecting the products available to

efforts are or should be designed to uncover any latent, subconscious needs that customers and consumers have.

Some Definitions

For clarity's sake some definitions of the terms used in the following are necessary. The problems and questions to be delineated for the new millennium can be seen best from a common perspective of understanding. Many of the issues to be resolved in the future require commonality of terms.

Neither the *Webster's New World Dictionary* nor the *Concise Oxford Dictionary* is of any assistance. Buying and selling figure largely in their interpretation of what marketing is. *Webster's* comes closest with its second meaning of marketing:

> *"All business activity Involved in the moving of goods from the producer to the consumer, including selling, advertising, packaging, etc."*

Much can be read into this and therein can be found the source of some of the confusion over the terms. Too much or too little can be understood from this definition.

Marketing

Marketing encompasses the broad activities in a company that include

■ Researching customer and consumer needs and desires for products
■ Translating this knowledge into reliable product concepts
■ Developing promotional and advertising material to convey to customers and consumers alike the availability and desirability of the product

Food technologists then will develop products based on these concepts and pretest them before they are launched. If these products are successful in test market — a joint collaboration between the marketing and sales departments — then they are turned over to the sales department.

To Stanton (1998) marketing conjures up two images in people's minds. These two images have a nasty tendency to merge, giving an almost

nevertheless catch one's eye and one remembers the product. For the average person these tools are most apparent in the stores where people shop, in the fliers they receive, and the TV program that are interrupted.

Stanton's other image of marketing is actually a philosophy of doing business, i.e., "making what the consumer wants to buy, rather than making the consumer buy what you want to make." This marketing philosophy emphasizes the consumer and puts foremost

■ The need to study consumers and customers (the gateway to consumers); to familiarize oneself with the needs and desires of consumers; and to develop products using the resources available or to develop the necessary resources that satisfy these desires.

Developing the tools (advertising and promotions) to communicate with and to attract customers is an important second function of marketing.

Marketing departments are responsible for investigating the needs of customers/consumers. From these studies they discover new markets for their existing products or uncover product ideas for new markets. They determine what their customers'/consumers' needs are based on their research; they draw up product concepts based on these perceived needs. After these concepts have been screened, technologists then can create products incorporating these needs.

In addition, marketing departments plan the strategies, the second function, that will communicate the benefits of their products to customers and they create the promotional materials for these products that attract consumers and customers alike.

This is more clearly seen in Figure 3.1. It shows the construct that is marketing. There are two major components:

1. Consumer and market research, that is, finding out who the consumer is and what the consumer wants and how to satisfy that want. Market research makes the company aware of developments in the markets and marketplaces it services. It exposes the competitors and their approaches to promotional materials. It helps bring an understanding of the customer to the food technologists working on product development and an appreciation of the competitive forces in the marketplace to the sales department.

2. The development and production, sometimes, of appropriate

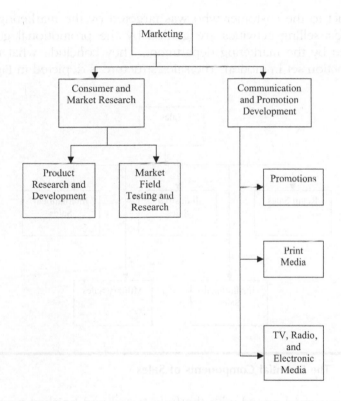

Figure 3.1   The Essential Parts of Marketing

Much of this creative work will be done with an advertising agency that has the skills and experience in all facets of media to produce material which attracts and interests customers.

If research, development, and test marketing go well, the sales department then takes over.

Sales and Selling

Selling involves interfacing with the customers (and retailers) in the selling arenas and using the promotional and advertising materials to advantage. The sales department also advises the production and distribution departments of their need for product in various locations as they prepare

the product to the customer who was targeted by the marketing department. Their selling activities are guided by the promotional guidelines established by the marketing department. They conclude what the marketing function set in motion. The sales structure is depicted in Figure 3.2.

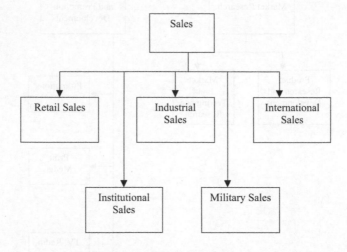

Figure 3.2   The Essential Components of Sales

Sales personnel armed with the fully tested and finished product and the tools designed to sell the product then proceed to the targeted customer in the various marketplaces where the market for the product is located. Each of these markets is or can be a complex of sub-markets.

Sales departments have a further responsibility. Sales personnel should communicate closely with marketing personnel to provide feedback about response to and acceptance of the product in the chosen marketplaces and how effective the communication vehicles for the product are. They should also report on any retaliatory action of the competition to the introduction of the new products so that counterstrategies may be formulated.

Markets

A market is an expression of an awareness of a need for something. That

organic foods, for kosher foods, etc.; that is, there are *demands* for these products. They are to be found in some marketplaces.

A market can be best described simply as a targeted food category for which there is a need or, conversely, a targeted customer for whom this need for a specific product exists. These targets are usually called niches. Table 2.1 classified some market niches according to the customer or consumer with some suggested product categories. Market niches could just as easily be sorted according to products. Each of these could be further subdivided and there is really no end to the niches that can be created.

The concept of a market is best typified by that old breakfast standby, orange juice. Canned and frozen orange juice first served a market niche as a breakfast drink. It was a convenience item to replace squeezing oranges. Then, with the addition of orange pulp to provide a source of fiber it moved into a new market, as a health food. More recently, cartons of fresh orange juice are cutting into the sales of canned and frozen orange juice. As such, orange juice with its added fiber became a refreshing drink that could be consumed on any occasion (actually representing a new market). With added calcium, a move was made to another market, a fortified beverage positioned against milk and aimed at both children and women, particularly those women worried about osteoporosis and who dislike milk or as adults have lost the habit of being milk drinkers. Doors were opened to new market niches and new targeted consumers.

Soft drinks made similar moves to new markets. From being simply refreshing beverages they were moved to a new market, as a respectable meal accompaniment. With vitamin fortification this move was solidified. Spritzers and wine coolers were developed to provide new markets (a refreshing drink without the seriousness of wine knowledge) while at the same time encouraging novices to wine drinking to become more interested in this beverage as a meal accompaniment. Even water was lifted into new markets, first into a health market as natural pure spring water, then with caffeine added into a health/energy/stimulant market. Salt, the most common seasoning ingredient, was moved into the gourmet market with such varieties as flavored salts (e.g., *gomasio*) and evaporated sea salts especially with those from some particular regions being highly prized. There are kosher salts and rapidly and slowly dissolving salts with their own specific markets. Within an ingredient market category there can be niches.

food manufacturers see opportunities for new markets and new products for these markets. Even space travel caused the development of its own unique market niche, albeit a very limited one.

The marketing department in the course of its research uncovers opportunities, i.e., areas such as the health food market, entrée items, sports foods, etc. into which a company's products can be put to target a particular segment of consumers. Not all markets are good for any particular company. The market must fit the philosophy, skill level, and technology of a company venturing into that particular area.

Marketplaces

Marketplaces are very simply where selling takes place. The nature of the marketplace varies according to the market (the target) and the targeted products being sold in them and also according to who the sellers are. Marketing gets the right buyers and the right sellers together in the right marketplace with the right products in the right market niche. Marketplaces will be discussed further in Chapter 4.

Food products must be marketed to be sold, and they must be sold in some marketplace if companies wish to prosper. Ergo, food market-places must contain the market whether that market is defined by the product category, i.e., snack food market, or by a consumer category, i.e. the teenager market. And both facts, the presence of the product in its market and the targeted market, must be made known in that marketplace.

The Technology of Marketing

The end of the millennium saw food products begin to have shorter and shorter life cycles. Companies demanded more and more innovation from the food technologists. Several factors may have driven this demand:

- Consumers, and therefore customers, were demanding greater variety and novelty as well as greater convenience, naturalness, flavor, and wholesomeness in the products they were buying.
- Consumers were becoming more fickle. That is, they were being subjected to so many stimuli that their attention was being diverted to other products and brands and to keep attention focused mar-

■ Improvements in market research techniques have given marketing
people better tools with which to obtain insight into the latent
needs and desires of consumers.

The marketing departments of food companies therefore must continually
update their arsenal of products and develop new ones to attract new
customers, keep regular customers, and satisfy the wants of all their
consumers.

A further demand is that these marketing departments of food companies must cope with developing new methods to communicate their
messages about their products to their targeted customers and consumers
alike. They must adjust to multimedia techniques with communicative
ideas that include TV for its visual impact; print for textual descriptions,
radio for sound impressions, and the Internet for its potential for information dispersal and retrieval and all this will be interactive.

Will better research techniques in marketing reduce the odds of failure
in the introduction of products or processes into new markets? In a perfect
world one *might* be tempted to say that the odds of failure would be
greatly reduced. Indeed, success might even be guaranteed in such a perfect
environment. However, on the outside of any food company's world, there
are two unknown factors: the competition and the general public (as
opposed to the targeted customer/client). The competition will certainly
react to any launch of a new product as a hostile act. In the perfect world
it is hoped that introducer of the new product had anticipated this with
its own plans for retaliatory action.

Some Imponderables to Marketing's Technologies

To do this requires new technologies to study the customer/consumer;
these were reviewed in the previous chapter and need not be rehashed
here. One point does emerge: There is a growing clamor against what
the public sees as a breach in privacy by marketing departments and the
research companies that they employ. The techniques to find out more
about customers/consumers are becoming invasive and very successful.
So much so that there is a rebellion against consumer researchers' hunt
for information about people's habits and a growing feeling on the part
of some consumer groups that customers/consumers are being manipulated.

including food companies, pay to have their products promoted under the guise of creating verisimilitude.

The public's reaction — and it must be remembered the public is a collection of customers/consumers — to messages about any new product or processing introduction is an imponderable. How will they react to the product's introduction? Certainly the introduction of irradiated products was greeted with outrage by anti-irradiation groups. Genetically modified crops have met with similar reactions especially from groups in Europe. Products from countries with poor civil rights records or even products with ingredients from such countries have been boycotted.

Even if the most modern techniques for customer/consumer research were conducted, there is no guarantee that the odds of product success can be reduced for a multitude of reasons:

- The research techniques may have been improperly applied in the wrong market or to the incorrectly targeted customer/consumer. If so, the results will obviously be incorrect.
- Results from properly applied tests may be misinterpreted. Such misinterpretation may be simple error or a bias introduced by a consumer research agency in order to placate and to retain a domineering client. Or a client's personnel may filter results to agree with the boss' known and determined views respecting a product. Without truthful and unbiased reporting of the marketing research, marketing, sales and R&D departments will make errors.
- Consumer and market research may have had nothing to do with a product's success. The food company's accounting department may demand an unreasonably quick return on the research and development investment and abort the introduction too quickly.

There is no assured avenue provided by market and consumer research to product development and introductions in new markets.

Marketing and the Global Village

connected and reachable. The phrase is becoming commonplace, indeed almost banal, in today's business-speak. It features prominently in headlines in business and trade newspapers and magazines. Speakers sprinkle their sales and marketing promotional talks liberally with this glib phrase.

It is a phrase that should mean something. Unfortunately, like much business-speak, nobody is really sure what it means but one likes its pithy sound. Nor do they understand its impact. Most people look at the global village as merely an opportunity for export sales. But export sales have been around for centuries, long before the global village was a catchword.

Webster's New World Dictionary, Third College Edition, defines "global village" as "the world regarded as a single community, as a result of mass media, rapid travel, etc." But a community suggests interests in common, a commonality of work or religion or political systems. This does not prevail at the present time. Certainly people have a common interest in eating but not necessarily eating the same foods.

Air travel, e-mail, and cellular phone technology have improved communication so that many now can visit virtually and converse in chat groups on a variety of topics with correspondents around the world. Satellite television can now allow receivers to view channels from around the world. How recently and how rapidly communication technology has advanced can be reviewed in Table 1.2.

But such mass media do not mean successful communication. Contacting someone anywhere in the world and, communicating with that person are two very different concepts. Communication implies understanding the message that was conveyed and acting in some manner on its content. Rapid communication and rapid transportation have not produced a greater familiarity that one usually expects to be encountered in a community atmosphere. As yet this is not possible.

"A world community can exist only with world communication, which means something more than extensive short-wave facilities scattered about the globe. It means common understanding, a common tradition, common ideas, and common ideals."

R. M. Hutchins

What Impact from a Global Village?

The new millennium will not see, for many decades, any change in this facile concept.

Ease of communication has not produced a better understanding of other food markets in foreign locations.

Global Marketing

The ability to sell frozen widgets in Europe, Asia, or Africa, i.e., internationally, is no premise on which to assume there is therefore a global marketplace. That one can sell frozen widgets worldwide is granted. In the middle of the past millennium goods were bought and sold globally or at least as far as the global world was known.

Problems arise for marketers, especially food marketers, when this concept of a global village is stretched grandiosely to phrases such as a global market or to global marketplaces. There is no global marketplace. It is a fanciful conception. At this stage, entering into the third millennium and with the tatters of the Seattle round of the World Trade Organization meetings (November/December 1999) still echoing, the difficulties and associated expenses to achieve this conceptual place far outweigh any advantages that might be foreseen in the new millennium.

But at this stage a global marketplace place is a collection of regional marketplaces with their own distinctive differences that are carefully guarded by nationalistic interests. This must be clearly understood and appreciated by marketing personnel.

Natural and Unnatural Barriers to the Global Markets

Why is there no global food marketplace? Five reasons are suggested:

1. There is no global food-based economy which would be necessary for global food markets and directly needed for global marketplaces.
2. There are no global tastes in food. Food tastes are based on cultural, religious, social, and traditional elements that are the very essence of the multi-ethnic nature of the world's population.
3. No company can administer its markets from a central location. Hence there will always be some degree of fragmentation respect-

5. Nations will insist that rather than submit to being exporters of raw materials for other nations to give added value to, they have processing performed locally.

A possible sixth reason militating against global food markets and marketplaces is the absence of any comprehensive global food policies.

Global Economies

At present at the beginning of the third millennium, world economies are in a turmoil. The World Bank has been criticized by needy nations for its harsh and frequently unreasonable (to the indebted nation) demands for economic reforms.

There are attempts at globalization of the economy. Some economists have suggested that there could be three regional economies: an American region based on the American dollar, an Asian region based on the Japanese yen, and the European region based on the Euro. Only European regionalisation has been initiated. Such regionalism has political, social, cultural, and economic problems to overcome.

Lind (1999) argues that regional integration, outside of the European Union, is not a trend and is, indeed, neither desirable nor inevitable. Regionalism, as exemplified by the European Union cannot be declared a success. There are some obvious omissions that one ought to expect in regionalism: There is no common foreign policy within the European Union and no common cultural identity. Therefore, regionalism is not inevitable.

If regionalism means regional, protectionist trading blocs, then global markets become a meaningless euphemism. "Nationalism-within-globalism, at the expense of regionalism, is the best outcome" (Lind, 1999). Sovereign nation-states will demand those differences of ethnicity and culture as represented by their foods and food habits and insist that the local food needs be attended to. France and Germany demand their food cultural identity.

Regional Foods — Yes; Global Foods — No

The second issue, a most costly and difficult issue for food manufacturers, centers on the products for global markets. There are very few added-

Corporation, to single out only one company, has changed its products to fit international markets:

> Roast pork on a bun with a garlic soy sauce is sold in Korea.
> In Indonesia, rice as well as french fries is sold.
> In India, mutton and vegetable products take the place of proscribed beef and pork.

Thus from a highly centralized philosophy regarding products, McDonald's Corporation has been forced to permit more local control and decentralization of marketing and decision making. Global marketing will mean servicing local customer and consumer needs with communication vehicles suited to local facilities, selling locally and building locally.

To devise new products to satisfy customer/consumer needs in regions around the world will require constant updating of customer/consumer information in each area, a vast resource base of suppliers, and a heavy expenditure in research and development. Development of new products to satisfy local and regional tastes worldwide will take its toll as a heavy, and costly, burden to food manufacturers. It would be ill-conceived strategically, as well as impossible tactically, to undertake such new product development from one central location as some multinationals have attempted unsuccessfully to do (Fuller, 1994, pp. 218, 219).

The need for regionalization of products and regional plants with their own development centers will prevail for many years into the new millennium. A food company's marketing research program that is driven by and from its centralized headquarters will very likely overlook important differences in customer or consumer behavior and competitive activity across national boundaries.

A corollary to this is the following respecting global markets: Food companies must use caution in taking their company or their products into any global market. Which marketing ploy is it wisest to use? Does a company use the company name? Or is it better to go with product names or a more encompassing range (of products) name or a regional name? There are advantages and disadvantages to each strategy. The company name approach certainly has promotional advantages over either of the other strategies. As Kraushar (1969) points out, promotions with the company name benefit the company franchise and thereby all the other brands and, if the name is highly respected, assists product entries into innovative but unfamiliar fields.

food products have a limited life. By associating the company name or the brand and the product too closely, a dying or passé product or product line can destroy a brand name.

Globalization Requires Company Fragmentation

This point has been alluded to from much of the preceding. Markets around the globe cannot be administered with a policy established in a central location. Any efforts to do so must be confined to the generation of policies based on generalities, i.e., produce only products with the highest quality raw materials, truth in advertising, best quality for the price, etc.

National policies, tastes, customs, and legislation will dictate how companies shall operate in foreign lands. One major retailer, Marks and Spencer™, found it could not profitably run its North American operations with policies, styles, and products developed and successful in England.

Lack of Food Law Agreement

Food products must conform to legislation and standards based on national food customs, eating habits, and religious restrictions for food products. This will require a very intimate knowledge of local and regional cultures, religions, and food legislation. Manufacturers wanting to send products into global marketplaces must respect food laws governing:

- The composition of products, particularly standardized products
- The packaging materials that are permitted to be used as well as labeling and ingredient and nutritional requirements
- The selling of products

All must be followed and all will influence costs.

Local Manufacturing Requirements

Finally, many nations will demand that a portion of the processing of local raw materials be undertaken in their country to support local labor needs. This is already a requirement in some countries. Some processing

that are producers of cacao, for example, require that some of the processing of the beans be done locally. Such requirements make it necessary to have processing facilities available locally or to have some local representative supervise co-pack operations in the region of manufacture and assess the availability and quality of local materials. That is, the global market concept requires attending to a series of regional markets with local manufacturing or contract packing facilities.

The reality is the following. Developing nations export their raw products to developed nations which allow the developing nation's raw unprocessed products in duty free. If that developing nation attempts any preprocessing of its raw material and thus provides jobs in its local economy, the developed nation imposes a duty on the semi-processed material. Should the developing nation produce the finished product and thus create more employment in its local economy, the developed nations raise the duty barriers even higher.

The Global Village Marketplace

Global marketing in a global village marketplace is a chimera. It will certainly not occur in the foreseeable future. There can be little anticipation of global marketing in a global marketplace until all the following issues can be addressed, that is:

- A global economy or strong regional economies
- Acceptance of diversities in culture, history, food habits
- Harmonization of food safety standards, food legislation, and nutritional criteria
- Agreement on (and assistance with) social, labor, and environment standards
- Establishment of an independent trade regulating body to rule on non-tariff trade barriers

Nor will it be entirely desirable that such globalization does occur according to some. Other barriers to global trade and hence to global marketplaces, i.e., non-tariff trade barriers, will occur. These will be more fully discussed in the section on food legislation

To market globally in today's so-called global marketplaces, food companies will be forced to build regionally. There are forces, largely

countered by a much stronger force pushing food companies to region-alization, to decentralization to better serve their regional markets and their customers/consumers and to accommodate local national govern-ment's economic plans.

Chapter 4

The Selling Fields:
The Marketing Arenas

"A market is a place set apart for men to deceive and get the better of one another."

Anacharsis

Setting the Stage

Anacharsis, a Greek philosopher who lived approximately 6 B.C., had a very cynical view of a marketplace, the subject of this chapter. The marketplace where goods are sold has changed dramatically. The caution, nevertheless, for the buyer to beware is always good advice in any transaction, but today's merchants need customers for repeat business in the very competitive marketplaces of the third millennium. In addition, customers and consumers are more educated and sophisticated. They now have become a domineering force, one that must be reckoned with by merchants and by manufacturers. Furthermore, governments in developed countries have legislation to protect customers and consumers alike.

The term "selling" will be used in a much broader context than conveyed by the term "retailing." Both selling (including wholesale,

communicating with customers. Selling carries out the logistical and tactical operations of marketing. Selling always operates with the guidance of marketing but not necessarily under its direct supervision. Selling is integrated closely with marketing in order to fulfill its tactical mandate and with manufacturing to carry out its logistical functions.

Simplistically, selling is getting the customer to buy the product. However, what might be simply described is a very complex issue with many sub-issues:

- A customer has been identified by the marketing department through its consumer research. This department has developed advertising and promotional tools to communicate with this identified customer.
- There is a product or service that this customer wants that must be displayed (advertised in some media), distributed, or retailed in a manner that attracts the customer and the consumer.
- There are costs to negotiate among seller (e.g., slotting fees), distributor, and manufacturer.
- There is a need to get the attention of this customer to convey that this product is available in some appropriate marketing arena.

In short, there are many steps between cup and lip that can complicate what is described as the simple act of selling.

The Many Selling Arenas

Selling normally occurs in some place casually referred to as a marketplace. This marketplace is the target of promotional materials and other attention-getting activities designed by marketing people with feedback from vendors and from the sales department who, as the tactical group, are interpreting the strategies of the marketing department.

It must be remembered that there are several arenas or marketplaces for food. There is a danger in looking at the many food marketplaces, the arenas for applying the stratagems of food marketing, as physical places. Some are not real, physical places; they are only conceptual places in the minds of the marketing department. For example, a seniors' marketplace is a marketing concept, it is not a real place. One cannot go to

■ The products are promoted in such a way as to appeal to seniors' lifestyles

Where is this seniors' food marketplace? Probably it is in the local supermarket but not garishly displayed as "food for toothless old fogies." Those food companies that have attempted to come out with, for example, soft foods for seniors (actually nothing more than gussied up toddlers food), have quickly died. It is interesting, therefore, to note that a supermarket can thus be subdivided into many niches.

There is also an ingredient marketplace. It is here that commodities and ingredients are bought and sold for further processing into finished products. The institutional or foodservice market (the catering trade, e.g., restaurant and hotel trade) is unique and can be considered as a separate marketplace. The marketplace for victualling the armed forces is distinct enough from the institutional market to warrant its special designation as a marketplace. Yet none of these has a real, physical location. There may be trade shows and exhibitions put on in certain venues on an annual basis but these, as marketplaces, are not physical places.

The Elements in a Simple Marketing Arena

Figure 4.1a introduces the various constituent parts of a simple marketplace (i.e., one without competitors or competing products) as a generalized concept. The four elements, the protagonists, that enter into the marketing arena are

1. The manufacturer. This can be the primary producer or harvester, the finished product processor, or the manufacturer of some ingredient to be used by a secondary manufacturer.
2. The seller who actually presents the product for sale to the customer. The seller can have many identities.
3. The customer or buyer. This is that entity which buys from the seller that which is needed for further sophistication or is offered for consumption. Again, the customer can have many identities.
4. The ultimate consumer or user of that which was purchased.

These, then, are the protagonists that comprise a simple marketplace. In Figure 4.1b, the complex interrelationships that must be understood

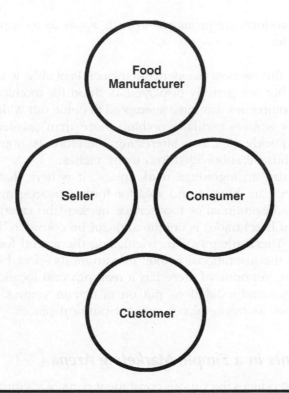

Figure 4.1a The Major Elements that Comprise the Marketplace

A. Seller/customer interaction where there is direct activity between the seller and customer. Promotional strategies developed by the marketing departments of food manufacturers and put into action by their sales departments will be used to attract customers. Retailers will have courteous staff to assist customers, well-designed layout of stores, and attractively displayed products to make shopping a pleasure.

B. Customer/consumer interactions that are defined by the needs and desires for products which were major factors used in the research and development of the products.

C. Consumer/manufacturer arena that is the dominant factor in new product development. Knowing what the consumer wants, knowing who the consumer is, knowing where that consumer is and how the consumer will use a product are necessary for product

manufacturer and the seller are. Sellers, especially the large chain stores, may have supply chain management or order management policies in place with manufacturers.

E. Customer/consumer/seller arena where most of the selling, as is typically recognized by the general public, takes place.

F. Customer/consumer/manufacturer interrelations where most of the consumer research takes place and promotional campaigns are developed.

G. Consumer/manufacturer/seller arena where media and promotional campaigns are tested and used.

H. Manufacturer/seller/customer arena where there are mutual efforts to attract, on the part of the manufacturer and the seller, and to be attracted to on the part of the customer; and finally

I. The main selling arena where all interactions are present.

Another level of complexity is introduced when all the elements in Figure 4.1b are laid against the backdrop of competitors.

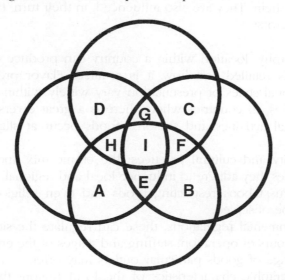

Figure 4.1b How the Major Elements Interact to Form the Various Marketplaces and their Relationships

The marketplace familiar to most people is the one where they shop

various ethnic groups, for vegetarians, foods prepared respecting religious rites, snacking foods, etc.

Selling, or retailing, takes place, then, in several different "marketplaces." Each involves different selling techniques; each involves vastly different customers (and consumers) who must be communicated with and whose needs must be ascertained. The overt trappings of marketing, i.e., advertising and promotion, used in one marketplace cannot always be used effectively in other market niches.

Problems of retailing become even more complex when each marketplace can have several subdivisions. The military marketplace has its unique requirements for foods; each service has its own special needs.

The Varied Marketplaces

The various marketplaces in which the selling of food or food ingredients takes place differ with the nature and preferences of the customers who are buying in them. They are also influenced, in their turn, by a multitude of external factors:

- Geography: location within a country can produce differences in what is retailed and how it is retailed. Flavor preferences and traditional styles of presentation vary widely within and between countries. In countries where there is a great diversity of terrain, seasonal activities and seasonal foods occur at slightly different times.
- Ethnicity and cultural practices: the ethnic mix and the cultural practices they adhere to influence local and regional marketplaces. Religious taboos respecting foods and even religious holy days must be observed.
- Governmental regulations: these can regulate the siting of stores, their hours of operation, staffing and wages of the employees, and the range of goods particular outlets may offer.
- Demographic characteristics of the local region: the age of the residents of the community, their level of prosperity, family size and age of the families and their shopping habits.
- Nature of the communities: whether the communities are residential or rural or light industrial or combinations of these will dictate the

firm statistics of how much product or produce is sold in the various channels be clearly defined.

The Traditional Outlets

What are traditional marketplaces in one area of a country may not be traditional in another area of that country nor even remotely known in another country. For example, in the city of Montreal in the province of Québec at the present time, the sale of cooked sausages from kiosks or mobile handcarts for eating on the run is prohibited. Yet 300 km down the road in Toronto in the province of Ontario such sales are permitted on the streets and the handcarts are a common sight. In the global village sense there are no universal traditional outlets. There are merely different types of outlets; some types are more common in one location than in another. Nevertheless, one would find that their operation, regardless of their location, would be influenced by many of the factors listed in the previous section.

Table 4.1 lists the more common traditional food outlets and presents examples of each. Worldwide, the volume of food sold in each of these outlets will vary from country to country as the selling of food is influenced by the development and maturity of the marketing-based infrastructure of a particular nation.

Some explanation of the general classifications is warranted. These are

- Craft-based outlets are crudely divided into two groups: farm based and skills based. Farmers sell their own produce (fruit, vegetables, eggs, cheeses, poultry, lamb, etc.) either in farmers' markets or at roadside stalls. The line between the two groups is blurred when many farmers and fishing families also provide added-value products such as smoked bacon; smoked turkeys; specialty fresh or dried sausages; cooked chicken or turkey pot pies; marinated wild garlic, cattails, and fiddleheads; smoked, marinated, and salted fish for local sale. To these skills-based outlets can be added specialty chocolate and candy makers, cheese makers, and bakers. They are (usually) owned and operated by a skilled craftsperson and (usually) sell only self-grown or self-produced product.

- Grocery stores. These are the primary outlets for selling food along

- Convenience stores are food stores which carry a very limited range of food items such as snacks, candies, cookies, milk, soft drinks, breads, and some packaged and canned goods such as cereals, meats, tuna, and salmon. They are usually owner-operated and cater to a local community.
- Non-food stores. Drugstores and department stores often sell food products. Drugstores may carry a range not unlike that of convenience stores (without a fresh meat or frozen food category). Department stores often have gourmet foods. In addition, some book stores and computer stores serve beverages and light snacks for browsers in order to encourage sales.
- Food service includes all catered meals whether they be eaten at home or away from home. These vary widely from the free choice market where one can order from a wide selection of items (e.g., restaurants) to the restricted choice market (catering as in train, bus or air travel, hospitals, nursing homes, school lunch programs) to vending machines. Some recent variations on this theme are prepared meals distributed to people in their homes (usually the elderly or incapacitated) and so-called chef rentals for busy families with no time to cook. In the latter, a chef shops, prepares meals in the client's home, freezes the meals, leaves a menu list of the meals that have been prepared and then cleans up.
- Industrial sales. This encompasses the sales of ingredients and other products between companies for the manufacture of either further processed products or finished products.

There are no clean separations between the marketplaces listed in Table 4.1. There are gray areas as one entity merges with another as they combine features of other outlets to become some hybrid food outlet. For example, supermarkets now have "big box" sections in their stores side by side with their regularly sized packaged goods. They also have in-store delicatessens retailing prepared entrée items, salads, and freshly pressed juices for take-out or consumption in the store.

At farmers' markets, retailers may be grower/producers who retail their own farm-grown produce in-season while off-season they regularly import produce from other countries. Thus they become a seasonal green grocer when their own produce is unavailable and a farmer-producer in season. Then, too, many restaurants sell popular menu items such as the chef's

Table 4.1 General Classification of Marketplaces Selling Food Products and Ingredients

General Classification	Examples
Artisan-based	• Farm gates, farmers' markets
	• Butchers, green grocers, bakeries, cheese shops, fishmongers, wine stores
	• Carriage trade, specialty and gourmet stores such as tea and coffee shops
Grocery store	• Family grocery stores, supermarket, box stores, bulk stores
Convenience store	• Paper/magazine stores, tobacconists, drugstores, department stores
Food service	• Full service restaurants, fast food restaurants, hotels, diners, happy hour bars
	• Take-out restaurants, mobile street vendors, kiosks, vending machines, cafeterias (work sites, schools), mobile canteens
	• Transportation meals
	• Nursing homes, hospitals, short-term care facilities, institutionalized care feeding, penal institutions
	• Personal chefs, rent-a-chef
	• Military feeding (officers' messes, cafeterias, field kitchens, combat rations, NCOs' messes, etc.)
Industrial arena	• Food trade exhibitions
	• Sponsored short courses demonstrating special ingredients
	• Sample distribution to industrial users

Placing Outlets: Improving the Odds of Success

"Location is everything" is a mantra repeated by marketing personnel everywhere. If a bricks-and-mortar establishment is to be erected it must be in a location that has a large body of potential customers in the

or might become unstable or are undergoing rapid transformation, or are likely to, will not prove apt locations.

The location must permit the type of activity and type of establishment that is planned. Thus, as much care goes into site selection as is taken in determining who one's customers are, where they live, and what needs they have so that new products can be developed to satisfy them.

The siting of any outlet where food is to be sold whether it be a grocery store, supermarket, or a dining or drinking outlet is a costly venture. The outlet must suit the selling arena in which it is to be situated. Finding such a site is a complex problem.

Sophisticated techniques to research appropriate sites are available. Davidson (1998) describes the technology used by the British brewer, Bass Taverns, to site an outlet. It is worth describing here as a preview of the technology for site selection that will be used and refined further in the new millennium.

Bass Taverns owns a range of brands as well as a range of restaurants. To site a restaurant properly with the correct range of brews presents a logistical problem. Brewers, therefore, study their clients well. They carefully analyze who the habitués are of the various classes of pubs they own. They are well aware of which brews are popular in the various types of outlets and what these regulars have in common. They know their average earnings, their spending habits (from demographic data), which drinks are most popular, and even how far the clients are willing to travel.

An example of a common breakdown made by Bass Taverns research is as follows Davidson (1998):

- First tasters are young, affluent, and are inclined to be adventurous types. They will explore new locations. Their tastes run to strong lager and stout.
- Blue-collar hunters are usually a home crowd who prefer a local pub with arcade games, a juke box, and bright lights. Their tipple is most often draught cider and ordinary lagers. They are generally unskilled laborers.
- Premium wanderers are drinkers of bottled lagers and stouts. They are true hunters, usually single and out to enjoy themselves.
- The pint-and-pension brigade are characterized as an older, limited income crowd out for a quiet sociable evening close to home.

- The cards and dominoes players want quiet. They often only want the social contact the pub provides and usually are locals.
- The nightclubbers are moneyed and looking for entertainment.

This is a formidable breakdown of customers/consumers yet as such is not unusual. A breakdown of a seniors' market described 18 categories and subcategories of seniors.

The brewer now had the task of finding a site where enough of the profiled customers they wish to attract will frequent.

The ease of obtaining demographic data from any number of sources has been discussed earlier. These data can be categorized geographically by information technology companies to sub-units of a couple of dozen households. With such informational maps, Bass Taverns can identify its patrons according to demographic profiling.

Working in conjunction with information technology companies and with a geographical information system, Bass Taverns developed algorithms that allowed it to identify areas where people met the profiles they had developed for its brands (more sophisticated than those above). Now the question became: Where on the map is the ideal location for the outlet with the greatest number meeting the profile? Some further criteria for the search might be

- A convenient distance to travel to the outlet should be determined
- A specified population density must be met
- A targeted average income must prevail in the area
- Unemployment must be low
- The age of potential customers in the area must be considered; most are to be below a given age range (with a minimum of pint-and-pensioners)
- Customers should spend above the average amount on going out to eat and drink

To oversimplify, likely and suitable sites for pubs can be chosen using very sophisticated geographical information systems analysis, a combination of fuzzy logic and search algorithms, (Davidson, 1998). (Fuzzy logic systems can analyze data that are approximations, i.e., data that do not quite meet fully the requirement specified above for a site by the brewer: or more simply put, can analyze data that are neither a 'yes' nor a 'no.')

Innovations in the Traditional Retailing Outlets

As the second millennium drew to a close the food retailing scene was chaotic. Gillmor (1998) described the changes taking place in supermarkets as Disneyesque.

There were detailed studies of how customers shopped and what bothered them as they did. Retailers wanted to know how shopping could be made a pleasant experience. Feedback from stores poured in about how customers hated the long checkout lines, how the elderly found the stores too big, and how parents with young children were none too happy with store layouts.

Changes were made. An open-air market concept with bright umbrellas suggestive of a European market was introduced. Food samplings became a regular addition. These demonstrated new products and how they could be used – and provided pleasant odors to tempt customers to buy. Aisles became wider (which also made stores larger). Other innovations were the introduction of in-store terminals where customers log in what their needs are and are provided with recipes or a map to find what they wanted. Some stores equipped shopping carts with scanners that allow customers to scan products themselves, obtain a total of their purchases and thereby reduce checkout lines (Gillmor, 1998). Everything was done to entice the customer to spend money in the high margin areas, the meat and produce sections.

Other innovations were the introduction of the big box stores to sell large sizes of goods (the institutional sizes) and as well as packaged and canned goods by the case. Prices are, of course, much cheaper when large sizes or large quantities are purchased which obviously proved a very popular feature with customers. Supermarkets met this challenge and featured this big box form of presentation of foods with box departments within their stores.

Grocery stores of all sizes challenged the fast food restaurants for the trade of the snackers or grazers, for the finger-food munching crowd, and for the no-time-for-breakfast crowd on the run. With coffee machines, hot chocolate drink vending machines, and juicers a variety of beverages can be dispensed for those on their way to work, stopping for a break, or just looking for refreshment perhaps for a thirsty child while shopping. It's a short step to also supply fresh doughnuts, bagels, or Danish pastries for these customers. And there was the introduction of in-store delicatessens.

Bakeries and delicatessens do similarly by providing espresso cafés for

backgrounds of the local people. It runs the gamut of poutine, to pizza, to hot dogs, to fish and chips (from a fishmonger), to Caribbean pasties, to empenadas to papas.

Loblaws™, a Canadian supermarket chain, has allied itself with the Swiss restaurant chain, Mövenpick™, to have a full supermarket with a buffet/deli style restaurant inside. Patrons can eat at tables, or munch as they shop or buy take-out for eating elsewhere. Juices are pressed freshly from the available produce. Bread, muffins, and coffee are all available. Some supermarkets even provide live music. This competitive gambit is directed at the fast food restaurant and formica-table-top diners.

The food retailing scene has truly become chaotic at the beginning of the third millennium.

This trend is not one that all approve of. Messenger (1998), for one, sees food retailers, the supermarkets in particular, forgetting what their core business is, that is, packaged foods and brand names. The possibility that supermarkets could steal away the ready-to-eat or the ready-to-heat meal section from the food service industry is ludicrous according to Messenger.

Other types of retail outlets besides food retail stores have been quick to use food to encourage customers who enter their premises to linger there, browse some of the core products that are on display and, of course, spend. Sales are encouraged in this more relaxing atmosphere over coffee and pastries. For example, bookstores now supply a variety of coffees, finger food of various sorts, and books, magazines, and newspapers for browsers. In a like vein, computer stores provide coffee, food, and e-mail facilities for their customers who can relax in a non-intimidating atmosphere. In all these instances, food is the key enticement to bring in customers; it may be sold and consumed but secondarily to the non-food items.

The Computer, The Internet, and The Marketplace

It was inevitable that these three should meet. The meeting has created not only new customer services but has created a whole new marketplace both for the institutional and industrial customers and for those gatekeepers, the customers. Initially this meeting was tentative but in the latter days of the old millennium, as more people possessed computers and had access to the Internet, the meeting has become a chaotic love in

It is the workhorse of e-commerce. But there is also the Intranet. This permits the rapid transmission of messages within a company whether that company is a single facility or many manufacturing, distribution, and sales facilities scattered around the world. It functions primarily as a means of internal communication for a company, allowing a head office to manage its satellite operations with real-time financial and inventory information, a centralized purchasing system, and instantaneous sales data.

Then there is the so-called Extranet. These are special links that a company may have with its clients or with companies with which it is joined in partnerships, cooperative transportation schemes, or co-pack operations. For example, a company's Extranet links would permit a client to trace the whereabouts of its order within the company. This service is especially important for just-in-time delivery systems.

Of most importance to the food microcosm in its relationships with clients are the Internet and the Extranet. These provide services for customers.

The impact of the interfacing of the computer, the Internet, and retailing merits a more general look than one focused solely on the food microcosm. E-commerce, the result of this interface, has changed and will continue to change the marketplace in several ways:

- Customers will become much better educated regarding the items they want to purchase. Manufacturers' Web sites can describe products in great detail with the result that customers are better informed. Food products can be pictured in use by cooks. Products such as cars can be virtually designed with all the options a customer wants and then viewed. Clothing can be viewed on models or even on the individual. Food processing equipment can be demonstrated in real operating situations.
- Customers will be able to comparison shop with greater ease by browsing various manufacturers' Web sites. Chat groups will provide comments on the experiences of others with the use of products, on special features offered by their manufacturers, and on services provided by the vendors. The Web sites of many clubs such as, for example, automobile clubs or consumer associations, already provide comparative information of costs, performance data and versatility of various products.
- Advertising and promotions, i.e., communicating with both cus-

Customers must register with a Web site, providing their name, address, a code name, and frequently other personal data including a credit card number and other financial information. As a result, customers tend to stay, i.e., stick, with those merchandisers with whom they are registered. Greater skills of communicating with customers who are dedicated to rival products or services (i.e., stuck) will be required to break this habit.

▪ The concept of a fixed selling price will undergo drastic rethinking; perhaps fixed selling prices for many products including food products in the wholesale market or virtual marketplaces will even become irrelevant. The more informed customer with a world of comparison shopping at his/her fingertips may opt for negotiated or auction pricing.

▪ The bricks-and-mortar store will not disappear. Merchandisers must, rather, eventually see and understand e-commerce as a complementary arm to traditional marketplaces. Conversely, e-marketeers must recognize the real (as opposed to virtual) services that the Main Street stores can provide. The advantage of on-line marketing over the old bricks-and-mortar marketing has yet to be crystallized.

This latter point deserves some further examination. There is much hyperbole written (see later, Keller, 1999) claiming that virtual companies with their Web sites will displace the Main Street store. Web sites are much cheaper to operate — after all, a Web page is cheaper than a Main Street store with all its associated overhead costs. This is not necessarily true. Operating costs for Web sites have risen steadily and can be expected to continue to rise in the young millennium. Web sites, the equipment to operate them, and the programs to support them must be constantly updated to keep a fresh image, to provide better services, and to remain competitive. Payrolls for operating Web sites are higher. Maintaining a Web site requires more highly skilled workers for its operation. The "hits" on a Web site must be regularly analyzed as any traditional marketplace analyzes its traffic. To use the site as a glorified electronic ordering tool is to not fully use the potential of the tool and ultimately to court financial disaster.

There are the costs for digital photography set-ups for a virtual catalogue (which must be constantly changed as new products become

In addition, the Main Street outlet provides a location where products can be viewed or returned for servicing, where defective goods can be returned conveniently for an exchange, or where customers can have their problems explained and solutions demonstrated. For many, e-commerce is too impersonal.

The operating costs of bricks-and-mortar establishments and those costs of the electronic machinery for a virtual marketplace must be compared and balanced carefully.

Further Views on E-Commerce

Keller (1999) comes to very different conclusions on what has emerged in the very short life of e-commerce's new marketplace. Three basic observations have become apparent:

1. Customers now have direct access to seller/manufacturers without a middleman as an intermediary.
2. Competition amongst sellers will be keen because it is very easy and cheap to set up a Web site. There is no need to spend money to build up an infrastructure or to have investment in bricks and mortar. But not everyone agrees with this (see previous section).
3. Pressure on profit margins will grow as everybody, seller and buyer alike, will know what is being charged for what quality of product. According to Keller, "It will be an economist's dream, a buyer's market, and a nightmare for many sellers."

These observations, too, deserve some comment. There is no argument that in some cases there was an attempt by some manufacturers to bypass the middleman (observation 1 above). But there was a backlash that might have been expected. The middlemen frequently did the selling and demonstrating of the product (in-store tastings, for example). They did the local and regional advertising. They also provided servicing of products and handled returns of bottles, aluminium containers, and spoiled goods. They were naturally annoyed at such treatment by their suppliers. The result has been promises (empty or not is yet to be seen in this very young market arena) on the part of manufacturers not to circumvent the resources provided by retailers.

That pricing will be shaken up goes without saying; "up" being the

of staff with marketing and selling skills and with some knowledge of the food products available. Again, the Web site must provide a service, give advice, be prepared to answer questions, and service problems that may arise with its customers. They must also attempt to sell complementary products. Such skilled staff for these tasks will come at a higher cost than is found in a Main Street store with its part-time workers (especially to be found in food supermarkets). The site cannot sell itself and cannot be looked upon as a grand order taker.

Processors will feel the pressure as they are forced into more and more broken shipment lots. Their shipping depots will be required to ship smaller lots to more destinations. They will be required to keep larger inventories on hand. Distribution costs must inevitably rise even if cooperative shipping systems are used.

There is fallout from this new way of conducting business. Increased trucking will be required for deliveries; consequently, there will be a greater number of trucks on the roads. Increased trucking has already been observed by many municipalities and other local governments. Truck-related accidents are increasing. The heavier traffic, both in volume and in weight, is damaging roads and, as a result, municipal taxes must rise to improve the road infrastructure. Trucking fees must increase and freight rates raised accordingly. All these costs must eventually be paid.

Finally, there is overhead cost. Operators of the virtual marketplaces get a percentage of the selling price of products from the local bricks-and-mortar retailers to whom they have referred their orders. This cost narrows the profit margin for the real retailer offset, it is hoped, by an increase in volume. Any loss must be made up in other areas.

Summary

The Internet created at the close of the past millennium another marketplace, a new one in which food business can be conducted. It will not, as some prophets have proclaimed, replace the traditional marketplaces. These marketplaces all have respected and stable niches. Certainly there will be changes as marketing departments gain more experience and maturity in the new medium of e-commerce. So far, merchandising by e-commerce operates in conjunction with traditional merchandising arenas. Perhaps it will merely become, in time, another traditional one. With time, too, the strategies and tactics adequate for the real marketplaces will

Enter the Electronic Food Marketplace

Some of the novel ways in which the electronic marketplace has been explored by food retailers to provide a broader range of services to their customers have been described by Hollingsworth (1997), Gillmor (1998), Krantz (1998) and Menzies (1998). The types of Internet service provided vary considerably:

- Customers at terminals within the store can log onto the store's databases. Here they can access information about what ingredients are required for the preparation of certain meals. They can be provided with recipes to use, a list of the ingredients that are required, plus a map of the store where the ingredients, produce and meat, fish, or poultry can be found.
- At terminals within the store, the database can offer menu selections to choose from if the customer is having a dinner party. Once the menu is chosen the recipes are supplied plus the shopping list for the number of guests. The customer is also prompted for such items as flowers, paper napkins, and suitable wines for the occasion.
- Nutritional information about the food products or of the recipes offered in the database is presented along with a cost per serving;
- Several of the larger supermarkets have Web sites on which people in their homes can visit the aisles, clicking on items they wish to purchase. After supplying payment information, the orders are then transferred to the closest store, filled, and delivered.
- Many Web sites for virtual grocery stores are actually store-sponsored with each store offering home delivery from its location (Hollingsworth, 1997). Other sites for virtual grocery stores maintain relationships with bricks-and-mortar grocery chains. In this manner the virtual grocery Web site can offer a shopping service with supply and distribution handled by a Main Street grocery store.
- Some companies provide on their Web sites only information about sales, promotions, or special events at their locations to surfers and do not offer online ordering services.
- Other grocers provide an ordering service but offer only a limited range of non-perishable items.

The computer and information technology coupled with in-store trained personnel can now supply the customer with an unparalleled

children or for elderly shoppers who find large food malls a tiring, uncomfortable shopping experience. By making shopping easier to select cheap yet nutritious foods, by demonstrating exciting menu opportunities for casual or special occasion dining with recipes provided, and by arranging the recipe items conveniently to encourage purchase of the additional items for the recipes, the shopping experience can be made more pleasant.

Some typical examples of e-tailers are IGA™ and Stong's Market™. IGA established Cybermarket™, an Internet shopping service in 1996. It began with 11 stores and grew to 149 stores by 1999. Groceries can be ordered at any time of day or night and picked up by the customer the next day or be delivered to the customer's home. After the customer has registered with Vancouver-based Stong's Market™, the customer visits, in cyberspace, whichever section of the store selecting whatever products are desired on the way. Ordering can be done by quantity, number or weight, and by brand name. As the order is accumulated, a total is calculated with each purchase made. Since shopping is often repetitive, i.e., with the same items purchased, Stong's Market also offers a Quick List of the customer's most frequently purchased items. Checkout is a simple click on "Checkout" whereupon the customer writes in the desired date and time for delivery (Menzies, 1998).

At present Stong's Market has two experienced store shoppers who fill the orders for the online customers who can also place requests respecting personal choice on fresh food products such as meats, fruits, and vegetables which the in-store shoppers endeavor to fill.

According to sources cited in Krantz (1998) online revenues for groceries in 1998 were $270 million. They are anticipated to reach into the billions within the first decade of the new millennium. Stong's Market reported that the value of online customers' purchases was on the average twice the value of in-store purchases (Menzies, 1998). The Food Marketing Institute predicted that eventually a third of current food retailers will provide cyberspace food shopping services (quoted in Hollingsworth, 1997).

These services, while they can be very sophisticated, nevertheless, are simply

- Advertising and information-providing vehicles
- Catalogues displaying food products from which customers can

book catalogue. It, too, allows one to search books by subject, author, or title and usually a brief abstract or certainly key words describing the subject matter.

E-commerce for food retailing has yet to mature to a point where there is a one-to-one relationship between merchandiser and customer. It is a convenience for those who do not have the time nor the inclination nor the desire to shop.

Changes to be Wrought by E-Commerce

For the new millennium food processors need to develop a new philosophy for marketing and for product development. Food processors cannot remain unaware of how e-commerce will change their business, its philosophies, and their relationships with their customers, the retailers. E-commerce has ushered out the era of mass-marketing. No longer is the effort made to provide and sell a product which was designed for a faceless customer/consumer who is representative of some targeted market niche to a mass market of these average buyers. That customer, armed with an electronic mouse, is in direct contact. With e-commerce the retailer must sell to an individual with a name, a personality, a real, live, flesh and blood individual!

In the e-commerce of the new millennium the objective of marketing will not be only to promote and sell *a product* to a customer but to sell a complementary line of products to a real individual. This will require skilled personnel (an added cost factor).

Furthermore, food processors will need to interact more regularly and routinely with virtual food retailers, those with no bricks and mortar warehouses or retail outlets. This will not be an entirely new phenomenon. Food processors have had to work with such retailers for a very long time. These are, of course, the specialty food catalogue retailers who have been around for ages. The pace of growth of such ephemeral retailers will increase with e-commerce.

Virtual food companies are not a new phenomenon. Many food companies exist in (brand) name only. That is, there is only a postal box address, everything pertaining to the business is outsourced. The products are co-packed by other manufacturers to specific standards. Quality control is assigned to outside professional laboratories. Sales and distribution are outsourced to agents. One prime example very familiar to me is a multitude

As the new millennium proceeds the rules of doing business in this new marketplace need to be clarified.

Point, Click, and Shop Selling

The principle behind point, click, and shop selling is simple. A customer types in a name or a description of what is wanted. The customer waits while the server searches its files to provide either sources of the product or sources of best fits. By browsing the various sites found a customer can comparison shop. Then with a click of the mouse on what is desired plus entering in the required proper identification and credit card information, the customer has made the purchases and then can sit back and await delivery of the item (see Krantz, 1998).

Certainly point, click-and-shop commerce has arrived for the food industry. It will, no doubt, continue to grow as more retailers are forced to compete using e-commerce. E commerce will not be a successful selling tool for all. At this very early stage in its growth, immediate impediments can be seen and while they are not insurmountable, they are imponderables:

- Not everyone has access to a computer and the Internet. Nor does everyone with access wish to purchase food items by e-commerce. Estimates of the number of households with access to the Internet vary widely and wildly from low figures of 20 to 30% to highs of 50 to 60%. The number of people with access is, however, growing rapidly regardless of whose estimates are believed.
- Not everyone will be willing to pay the delivery fees for the service – up to $10 per order — so middle and lower income customers may stay with the traditional outlets. E-commerce is a benefit for those people who can afford the extra cost and consider it worth the time they save.
- There are concerns about safety of credit information and of privacy respecting the transfer of any information electronically. Many customers are fearful of the amount of personal information about the habits of e-users that becomes public to the merchandiser.
- There is distrust of the truthfulness of "infomercials" on the Internet. At present there is no regulation of these advertising tools that Web surfers can encounter.
- The customers' reluctance to entrust shopping to others has yet to

customer to make last minute changes or make substitutions for products that are out of stock.

Will profit margins for food retailers be squeezed by e-commerce? Will fixed pricing for food items die as customers have the ability to surf several virtual stores and comparison shop (see Keller, 1999, page 22)?

Influence of E-commerce on Products, Suppliers, and Customers/Consumers

E-commerce will without a doubt have an impact on food production, products, and services, on their suppliers, and on customers and consumers, the recipients of these products and services. At present, how far e-commerce will progress in popularity with the general public is anyone's guess.

A most probable observation is that business, and the food business in particular, will use e-commerce in more sophisticated ways for, for example, such activities as electronic data interchange or efficient consumer response. What is yet to be resolved is whether the power of the Internet, e-commerce, and associated business tools will result in the segregation of the have-companies and the have-not-companies; i.e., companies with the resources to develop these systems will grow ever bigger and dominate marketplaces while small companies which lack the equipment, technical staff, and financial resources to utilize these tools will suffer.

A clean separation of the effects that e-commerce will have on the food microcosm is not possible. Each link in the food chain has some influence on its neighbors.

Products and Services

Web sites can be used to educate customers about products and services. Food ingredient suppliers, through their Web sites, can educate manufacturers of added-value products and inform about the uses and properties of the ingredients that they have. As a service feature for their customers, technical staff can answer specific questions concerning ingredient applications online and in real time.

ingredients and additives to provide the desirable characteristics in foods that consumers want can be explained. As a two-way communication tool it may allow the food manufacturers to read what the concerns of their customers are and, by understanding what they are, correct them.

A more obvious service for food retailers is the placement of advertisements on Web sites, to replace the fliers that are sent door-to-door. A disadvantage is the inability to provide a couponing service on the Internet to promote a local store or special sales reductions of particular items.

Suppliers, Manufacturers, and Middlemen

Manufacturers and retailers have always worked hand-in-hand. The success of one has depended in large measure on the other. E-commerce can change this cooperative relationship.

Retailers with their bricks-and-mortar retailing establishments (often sitting on prime and expensive real estate) are coming into conflict with their suppliers/manufacturers (Tedeschi, 1999). For example, retailers of non-food items often provide customers in their stores with information and demonstrations but then may not get the sale as the customer goes to the supplier's Web site to take advantage of any discounts given by the elimination of the middleman, the retailer. These savings can be appreciable especially for big ticket items. Sales are stolen from bricks-and-mortar retailers. They did all the selling and demonstrating to potential customers. They lose to the virtual retailers or to the manufacturer operating a Web site where products can be sold.

This is a cause for tension between the Main Street retailer and the e-commerce retailer. This situation may certainly be the case for some big ticket non-food items. Would it be possible for electronic shoppers to bypass other intermediaries? Such practices are becoming more and more common for lower priced goods such as clothing, shoes, and cosmetics.

Food sellers in the new millennium are becoming a blurred entity with the advent of e-commerce. Exactly who are they?

■ Supermarket grocery chains that have added a Web site. From this, customers can order food items for pick-up at a local outlet owned by the chain or for delivery to their homes at some convenient time.

■ Traditional outlets in malls or stand-alone stores on main streets.

- Virtual outlets that exist only as a Web site. They may maintain a distribution center from which orders are shipped out to customers;
- Discounters that are variants of the above. This virtual grocer subcontracts with a large supermarket chain which fills the orders given to it by the virtual grocer from its many stores to customers locally.

Sellers now have four media in which to advertise their products and promote their services:

1. The print medium with newspaper advertisements and the traditional hand-delivered flier. Both will always be effective because they can be used to promote locally available produce or specials. They can also be a vehicle for any coupons or piggy-back offers.
2. Local radio is a vehicle to advertise food items and promote the availability of an Internet service.
3. Local television can present the pleasant features, conveniences, and products of a supermarket.
4. The Internet itself is a medium for advertising and promotions. Attractive and informative Web sites with their infomercials can be effectively used to attract, entertain, and inform customers.

Food processors require new food product development to replace aging products that customer/consumers no longer want. Development and introduction of these are costly ventures. Part of the cost of a new product introduction is slotting fees charged by stores and chains. If a manufacturer wants a new product to get shelf space, then stores require a fee for providing it. Stores require that each foot of space must return a certain yield to the store. Products that do not produce that return get dropped. Consequently, the introduction of a new and untried product might promise to lower the shelf yield. Thus does the store justify the slotting fee. It is their insurance against facing a possible loss of income.

E-commerce introduces the interesting possibility that manufacturers could bypass the slotting fee demanded by retailers. The manufacturer, by using its Web site to advertise the new product and develop a demand for it, could circumvent the middlemen, the retailers, and ship directly to Web customers. When demand does increase, retailers *might* find the manufacturer unwilling to pay a slotting fee. Only time will tell whether

The Selling Fields: The Marketing Arenas ■ **111**

Customers and Consumers

It was only logical that market researchers would be curious to determine:

- Who were using e-commerce
- How and for what were they using it
- What were their concerns in using it

U.S. statistics provided by Krantz (1998) found that the users were young, mostly single, and mostly with a college degree. Groceries were the third item by dollar volume after travel and computer hardware. They feared hackers and the need to divulge personal information (this latter topic will be discussed in a later section).

The customer has been empowered further through the Internet. It has given the customer the ability to navigate a multitude of Web sites of suppliers of competing products where comparisons and evaluations can be made. This has been made much simpler for customers by the use of "bots."

Bots joins a list of esoteric and arcane names and acronyms loved by computer-knowledgeable people to befuddle others less in the know. It stands for "robots" that can roam any number of virtual stores as virtual robots (to continue the analogy of robots), using various search parameters to obtain specific information about articles for sale. Bots are large Web sites called portals and use what are basically data-mining techniques to explore many sites. From these sites items based on the search criteria the customer-browser is interested in can be pulled out and displayed for the customer-browser. Price is obviously one search criterion that would be used. A customer would be presented with an array of sources and prices for particular items. By pointing and clicking the item is pictured and clicking on the "buy" button puts it into the shopping cart.

No longer do industrial customers, for example, have to search in their files through a number of usually out-of-date catalogues for suppliers to satisfy their ingredient or equipment needs. Now they can access a wide variety of suppliers of ingredients, equipment, and raw materials. They can get up-to-date information respecting standards and specifications of all their material requirements. They can converse, via e-mail, with experts for advice and obtain answers to specific questions. In short, they become better informed customers whether they are at the retail level or at the commercial level.

- Customers can converse with others about products they have used. They can describe how they have used them and provide recipes for others using the products.
- Customers can discuss deficiencies they have found with products they have used.
- They can evaluate services they have received regarding equipment; and above all,
- They can compare prices that others have been charged.

A search for recipes for dog biscuits had me inundated with recipes and process details for making them from veterinarians, kennel owners, and breeders.

A real danger for customers/browsers is the skewed information that can be available disguised as genuine. This can be true of information found in chat groups. The novelty of the Internet has not yet allowed the development of a healthy skepticism in all browsers about what they see written there. With the long history of the printed page, people (some of them) have developed a critical eye with which to sift out nonsense from truth in books. People have, for example, learned that diet books that promise extraordinary weight losses in a month should be read warily. Time will sharpen people's acuity in separating fact from fiction on the Internet. Until they do develop this discrimination, there is still not a deep trust of information found on the Internet. Customers are leery of the origin, and veracity, of the information they receive. The promotional statements about health(y) food products, i.e., some of those with added nutraceutical ingredients, can be false, misleading, or even provide dangerous misinformation.

One of the advantages for merchandisers selling via the Internet has been stickiness (see earlier). Once a customer is registered with his/her name and credit card recorded it is very easy for customers so registered to just click and buy. Customers, so to speak, have been captured by the ease of the system. This stickiness is not always to the advantage of customers since they, by browsing or by using a portal, might have been able to obtain better food buys elsewhere.

Security in the Electronic Marketplace

The damage that hackers have caused on the Internet is well described

2326/ch04/frame Page 113 Thursday, December 14, 2000 2:59 PM

using the Internet for shopping, fearful for the security of personal information that they have provided, information that may include their credit card numbers or their bank account numbers. In short, there is a lack of privacy and security from hackers.

Security, privacy, and price have been recurring themes in the initial stages of Internet commerce. The two greatest fears of customers using the Internet for e-commerce are hackers and the need to reveal personal information to register for a purchase (data reported in Krantz, 1998). These fears have yet to be allayed. Until they are, e-commerce cannot become a dominant marketplace.

Sellers on the Internet require registration of their customers. They also frequently ask for other personal information always in the guise of "hoping to serve you better." Customers are understandably reluctant to provide too much information. For sellers, the next research objective is to use sophisticated filtering and data mining technology to study the patterns of their customers' purchases to get an idea of the nature, personality, and lifestyle of these purchasers.

With this information in their databases and as their customers make further e-purchases there will be an accumulation of data to profile specific customers. Data mining technology permits suggestions of other products both food and non-food items that complement the previous purchases. A danger for the customer comes when these links to other sites, i.e., other products and services, are not declared as paid links or independently identified links. The customer can be deluged with unwanted services.

A Recapitulation

A great amount of space has been devoted to that selling arena described best under the blanket term of e-commerce. No special importance should be placed on this space allocation other than that this is the newest selling arena and is, according to all media reports and statistical studies, the fastest growing segment. It is also an arena that is not yet clearly understood by the food microcosm. It is still a novelty that many in the food microcosm have not yet taken seriously but will in the new millennium be forced to concede some of their marketing, selling, and buying activities to. Therefore, it is wise to review some of the results of e-commerce in non-food fields.

understanding of customers' food buying habits which can then be correlated with their profiles. It is a boon to customer/consumer research. It also represents a danger to privacy which is not acceptable to many individuals.

It is expensive, but many merchandisers as well as food processors may be forced into some aspect of e-commerce for competitive reasons. They will need it to maintain brand awareness, to advertise new products and services, and to defend against false, malicious, or even deliberately misleading information spread in chat groups and personal interest Web sites.

That area of marketing involving the design of advertising and promotional materials has been dramatically changed by e-commerce. Specifically, the design of Web sites has become a new advertising medium. Marketing specialists will have to rethink and redesign their strategies for communicating with customers; sales personnel will have to adopt and adapt another selling tool. Web advertising technology has enjoyed one advance beyond the banner advertisements upon which browsers clicked and then were transported to another Web site. Now "live" banners permit browsers to click, see an expanded message, and not leave the original Web site — a technology that has allowed more responses to live banners than to "click-through" banners.

Portals, or the more picturesque bots, which combine a search technology with data mining technology give promise of permitting the development of comparison shopping to levels never before conceived. What this will do to prices when applied to the food microcosm is anyone's guess. Lower prices would seem at this early stage of e-commerce to be the logical result. This has even resulted in one wag on the radio recently commenting that he had an excellent money-making idea. He would give products away and make his profit on the handling and mailing charges! He was belaboring the experience he had when he purchased some books on the Internet; the handling and mailing charges were greater than the cost of his books.

The conclusion that there will be a lowering of prices conflicts sharply with the opinions of others who see only increasing costs for customers with e-commerce.

How e-commerce will emerge during the early years of the millennium will become clearer when the hyperbole surrounding it has been blown away or been tempered by the experience of all elements in the chain from primary producers to customers.

but did not displace farmers' markets; supermarkets in their heyday have not destroyed the corner grocery store. The super malls grew as a challenge to all but have not displaced any of them. Each has been world famous for fifteen minutes. E-commerce will also have its turn.

Chapter 5

Nutrition and Health

"Eat to live, and not live to eat."

Benjamin Franklin

"...the investigation of the science of nutrition – a subject so curious in itself, and so highly interesting to mankind, that it is truly astonishing it should have been so long neglected..."

Sir Benjamin Thompson (Count Rumford) quoted in But the Crackling Is Superb, edited by N. and G. Kurti, Adam Hilger publishers, Bristol

A Growing Concern

Knowledge that food is good for you is not new. Early humans certainly knew that the absence of food for long periods of time meant starvation and ultimately death. A successful hunt or a plentiful harvest determined whether the community survived periods of scarcity or not. That much knowledge prevailed before there was awareness of the importance of the macronutrients, the micronutrients, and now the biologically active non-nutrients which have some effect on the health of the human body.

Prevalent diseases that the early population suffered are recorded by

1957) (see also, Table 1.3). Life expectancy (1340) was such that few people reached more than 40 years of age and fewer still lived beyond 50 years of age (Durant, 1957).

What was not generally known in these early centuries was that a particular quality of food, its nutritional value, was an important factor in both health and longevity. They were not aware that food contained essential factors for the normal, healthy functioning of the body. That knowledge is a comparatively recent awareness in European and North American cultures.

Lind (see Table 1.3; Singer, 1954), for example, published his study on scurvy only about 250 years ago citing its cause as the lack of fresh fruits and vegetables thereby providing the cure, but not the curative factor, for scurvy. More than a hundred years later, Takaki (Itokawa, 1976) established food as a factor in beriberi but again the causative component remained unknown. Hopkins (Hawthorn, 1980) at the turn of the 20th century produced the first report on that important group of micronutrients, the vitamins. Then, in fairly rapid succession the many different vitamins, important co-factors (such as the minerals), and vitamin-like components of food were discovered. The role and importance of protein and amino acids in the diet were elucidated. The value of fat and the essential fatty acids in the healthy maintenance of the body was revealed. Nutritional knowledge of the value of both the macro- and micronutrients has grown enormously since the first vitamin was isolated and identified.

That the prevention of diseases such as cancer and heart disease could be related to the consumption of a variety of foods variously called functional foods, nutraceuticals, pharmafoods, nutricines, or medifoods, or to factors which are found in such foods, is an even more recent development. It is one branch of nutrition that is undergoing intense study and unfortunately vigorous exploitation by scientists, by the food industry, and by the media.

The claims for such foods are that when they are a significant part of the diet they promote good health, strengthen the body's immune system and consequently contribute to the body's ability to fight diseases. Fiber was one of the earliest such components touted. It was alleged to reduce cholesterol in the blood and to play a role in the prevention of colon cancers and some other cancers. Food manufacturers were quick to jump on the bandwagon and have used fiber, extracted from whatever previously waste plant byproducts that might contain fiber, to develop new

areas. The result has been a labeling nightmare as some herbal folk medicines become mainstream ingredients used to fortify many products from breakfast cereals to beverages.

The Nutritional Sciences

Nutrition's role in the proper functioning of healthy bodies and in the prevention of nutritional deficiencies is now appreciated and recognized. The mechanisms of how this is accomplished, however, are not completely understood. Scientists, for example, do not agree on how some of the micronutrients create the reactions that are attributed to them.

The role of these components in the prevention of specific diseases is not at all well understood. Nor is it clearly established that all claims made for these foods containing these components are true.

The subject area of greatest interest to nutritionists is now the elucidation of mechanisms by which diet is a factor in one's well-being. They are now concerned with:

- The importance of nutrition in the performance of human beings under extreme stress as in manned space flight, in military operations, in post-operative recuperation (enteral foods), in radiation therapy, intense athletic endeavors, etc.
- How diet and one's genetic make-up interact to influence one's health. An individual's genetic make-up may greatly influence absorption, metabolism, and gene expression.
- The action of non-nutrients in shaping and influencing the body's defense mechanisms or its physiological or mental state.

Nutritionists are now focusing on preventive nutrition. Food manufacturers are pushing products that can capture the marketing potential that preventive nutrition represents. When the ten leading causes of death are believed to be either diet related or to have diet play some role in their etiology (Belem, 1999), then nutrified products certainly have market potential.

In fact, the market potential for such health(y) foods is enormous (Belem, 1999):

■ Japan's market is estimated at $60 billion per year.

Preventive maintenance foods, therefore, represent an important research area for nutritionists and for food technologists for developing new products to satisfy the needs of the public in the third millennium. Dietary regimens directed toward preventive nutrition will inevitably lead to what already have been called "designer foods." These are products formulated with biologically active non-nutrients and designed to prevent (a more accurate term would be "to play a role against") specific diseases or to maintain a healthy lifestyle.

The Macronutrients

Proteins, fats, and carbohydrates are important both quantitatively in the diet and as carriers for the essential micronutrients. Due largely to caloric content, overabundance and underabundance of the macronutrients are concerns for people's health; too many calories lead to obesity in the individual and too few calories result in under-nourishment. Both have serious health implications.

The quality of the macronutrients is important. A complete complement of the essential amino acids, the basic building blocks of proteins, is needed. Animal proteins are considered to be higher quality in this respect than are vegetable and cereal proteins. Intense research in cross-breeding and genetic engineering is being conducted to increase both the protein content of cereals and the quality (i.e., the completeness) of the protein contained therein.

Dietary Fats

The public has been encouraged to reduce the amount of fat in the diet such that 30% or fewer of the total calories come from fat. This reduction, in large measure, has been accomplished. Paradoxically this has not reduced the number of overweight or obese people in developed countries. Now there is suggestive evidence of a virus that may cause some people to gain weight (Holmes, 2000).

From the amount of fats or the amount of calories due to fat in the diet, concern now has switched to the kinds of fat that are consumed.

hit against the amount of animal and saturated vegetable fats (palm oil and coconut oils) as major components in the total fats consumed.

There emerged the physiological need for the essential fatty acids in the diet. They were implicated in prostaglandin synthesis and hence disease resistance and in heart problems.

The consequences are somewhat of a dilemma for consumers. For good health, calories derived from fat should be kept low. But fats are a vehicle for the fat-soluble vitamins and their component fatty acids are essential in the diet. The essential fatty acids are the mono- and di-unsaturated acids. Thus, the consumer is asked to regulate the amount of fat intake while maintaining an adequate supply of fat-soluble vitamins and essential fatty acids. Now it is strongly suggested that reducing the amount of fats in the diet is not necessary. Eating the right balance of ω-3 and ω-6 containing fats is more important (Neff, 1998).

Obesity and Disease

The fleshy female figures of earlier years as depicted by Rembrandt's Bathsheba or by Goya's voluptuous Naked Maja, or abounding in a Rubens picture are definitely out in today's fashions. Thin is in and fat is out. Or is it? A walk along any city street in a North American city will reveal that a very large percentage of one's fellow citizens are not just fat but obese. While fashions may come and go respecting the shape of the human body, the rapidly accumulating scientific knowledge backed by statistical studies on large populations indicates that the overweight condition is unhealthy.

And expensive. Birmingham et al. (1999) studied the direct health care costs of obesity-related illnesses (cardiovascular disease, hypertension, diabetes, and certain cancers) in Canada in 1997. Coronary disease accounted for 346×10^6, hypertension 656.6×10^6, and obesity-related diabetes 423.2×10^6 in 1997 alone. At that, the researchers consider this figure to be too low since diseases having multiple causes not linked solely to obesity were not included in their study.

The health and social problems associated with obesity are detailed more comprehensively by Lachance (1994). In addition to the cost element of obesity (Birmingham et al., 1999), obesity is a factor in

■ Gout

There are also social and psychological problems associated with obesity: discrimination and prejudice against individuals because of their obesity. Obese or simply overweight individuals feel stressed in a culture that values slimness. Such stress can foster in its turn a host of psychological problems in some individuals. Antipathy toward obese individuals can often lead to reduced economic capacity of the obese population. Employers are reluctant to hire the obese because of their proneness to accidents, their inability to perform certain tasks, and their greater incidence of illness causing absenteeism.

The soaring costs of obesity, both direct and indirect, have attracted the attention of governments. The World Health Organization, for example, holds obesity as a serious worldwide epidemic. A Reuter's news dispatch quoted Dr. Philip James, chairman of a WHO task force on obesity at a congress held in Spain (Anon., 1996) as saying,

> *"Obesity is doubling every five years so we have an epidemic that is coming at the health service like a tidal wave… a disaster on our hands."*

Obesity is on the rise in North America (Neusner, 1999) despite a proliferation of food products that are low fat, contain no fat (using fat mimetics), or are reformulated using synthetic fats (fats which are not absorbable by the body but have fatty organoleptic characteristics), for example:

- Car seat manufacturers have had to make seats for trucks, large cars, and sports-utility vehicles larger. Airplane seats are being made with higher drop-down trays that do not catch the bellies of larger passengers. More seat belt extensions are required on flights.
- Government surveys show that 50% of U.S. adults are overweight and 33% are actually obese. Approximately 50% of women in the U.S. wear clothing size 14 or larger. The average daily caloric intake is up for all ages of men and women since 1990, 11 and 6%, respectively. In Canada, approximately 33% of adults are categorized as obese (Birmingham et al., 1999).

Statistics on the overweight condition and definitions for obesity vary widely as there is no universally accepted definition of either overweight or of obesity. Lachance (1994) discusses some of the methods of mea-

public must adopt good eating habits. These can only be inculcated into the mature public through education in food, nutrition, and health while people are youngsters.

Where obesity is a problem and medical costs have soared as a result of health problems caused by obesity, governments of developed countries have reacted to the problem of obesity in various ways. They have:

- Introduced nutritional labeling of food products to guide consumers in their purchases;
- Relaxed laws regulating the health claims that food manufacturers can make for their reduced calorie products or designer foods;
- Ruled on the safety of non-nutritive intense sweeteners or fat mimetics, extenders, or synthetic fat-like substitutes as additives; and, on a more positive note,
- Promulgated and promoted new food guidelines for consumers.

Some vested interest groups have been highly critical of many of the recommendations; for example, decreasing consumption of animal proteins in favor of more fruits and vegetables has annoyed the cattle ranchers.

The success of these measures must be measured by the general growth of the population of the developed world:

- Fatter rear ends
- Soaring costs of more diseases directly attributable to the overweight condition
- Increasing consumption of calories combined with a more sedentary lifestyle

In short, these efforts have been rather disappointing.

At the same time that governments published their nutritional guidelines and required nutritional labeling, they were reducing grants to health education and other social programs which could have complemented these efforts.

The food industry responded to the problem of obesity with an abundance of low-calorie, low-fat products. The products ranged from low-calorie beverages to low-fat hamburgers. Farmers were encouraged to raise leaner animals.

Food scientists found more areas of research as they worked to develop new products or ingredients to curb appetite, or magically dissolve fat than to educate consumers and to promote a healthy lifestyle.

A Backlash?

How well have these measures worked? Apparently, after stupendous growth, sales of low-fat or low-calorie foods are either stagnant or lackluster. The diet cola market, while still a big U.S. market at $13 billion a year, has not grown substantially in 7 years. Obesity is still rampant. Obviously fat-filled food products must do better in the marketplace than fat-free or low-calorie foods in the marketplace or (*vide infra*) people are eating more of the low-calorie, low-fat foods in the false belief they confer no calories.

Neusner (1999) reports that manufacturers are responding — indeed, have been forced economically to respond — to a trend toward full-fat products (!):

- Nabisco Holdings Corp. added fat to its Snackwell's™ line of low-calorie products with the result that the falling sales of the product line stabilized.
- Ben & Jerry's Homemade Inc., a well-known North American ice cream maker, saw their full-fat ice cream sales rise 29% in its last quarter in 1999.
- Olestra™, a synthetic fat substitute from the Procter & Gamble Co., has had little impact on snack sales. Its promise has fizzled out.
- Taco Bell™ withdrew a line of low-fat fast food products it had developed. Their main consumers were teenaged boys who were not interested in fat-reduced foods.

And consumers are still not nutritionally aware or knowledgeable. Labeling and nutritional guidelines have not had a major impact on eating habits. If they had, obesity would not be the problem it still is at this, the beginning of the third millennium. The following anecdote illustrates the problem regarding nutritionally related disease:

> *A biochemist, a professor at a major university, commented to me on his passion for snacks, corn puffs, and corn and potato chips in particular. He was delighted with the coming of snacks prepared with*

Herein is the problem. Consumers, ignorant of even a basic awareness of nutrition, when confronted with an array of low-fat or no-fat products, believe that they can consume large quantities with impunity. The erroneous assumption that is made in their minds is that now they can have two pieces instead of just one!

When adults, who should know better, are so ignorant of food and nutrition and when young people are growing up with little or no food science or nutrition training or cooking experience, bad eating habits leading to nutritionally based disorders are to be expected.

Proteins and Carbohydrates

One is not too certain what to say about proteins and carbohydrates in nutrition and health. Fat alone cannot be blamed for the conditions of overweight and obesity and the problems associated with them. Both proteins and carbohydrates contribute to an overabundance of calories in the diet leading to obesity.

It is unfortunate that protein nutrition has been rather eclipsed by the attention given to its two sister macronutrients: fats and carbohydrates. Because of their high caloric density, fats are the focus of attention for attacking the problem by manufacturing low-calorie foods. Carbohydrates get attention for two reasons:

1. First, the simpler carbohydrates contribute sweetness and consumers have a sweet tooth. Hence consumers desire sweet foods; consequently, carbohydrates contribute significantly to caloric intake. The search is on for non-caloric sweeteners and non-caloric fats with which to make low-calorie food products with sweetness and with the satiety feeling contributed by fat.
2. Second, the more complex carbohydrates, the polycarbohydrates, appear to play some role as biologically active non-nutrients.

It is because of the second function that carbohydrates have garnered intense interest both by nutritionists and by manufacturers of ingredients and finished products. Fiber, a generic term for a complex polycarbohydrate, has become an important food ingredient as a biologically active material playing a role in the body's disease defense mechanisms.

Proteins have had no such fanfare to cause them to be the center of

and its complement of essential amino acids. Both conventional and gene transfer techniques are being used to accomplish this enrichment.

Of some unusual interest is the amino acid L-theanine. This amino acid is found very sparsely distributed in nature but is found in comparatively large quantities in the tea plant (Juneja et al., 1999). It is found as a free amino acid; that is, it has not been found combined in protein. Its physiological action as a relaxant is explored in depth by Juneja et al. (1999). Already a commercial application, Suntheanine™, has been marketed.

The greatest concern respecting protein is not nutritional but economic and social. As developing countries become more affluent, there is a strong desire for their peoples to increase the amount of animal protein in their diets. Vegetarian diets do tend to blandness unless spiced. Animal protein, even in small quantities, provides a welcome flavorful body to such dishes.

Wherein, then, is the problem to be faced in the third millennium? The trend to greater consumption of animal protein puts pressure on the production end of the food microcosm. Already there are concerns about the dwindling stock of seafood. Political efforts are being made to get agreement between nations to stop or curb over-fishing. Environmentalists and moralists are concerned at the loss of forested land to grazing land for animals and point excitedly that it takes more, several pounds more, cereal grains fed to animals to raise one pound of meat. These grains could support more people than the one pound of meat. This issue will be discussed more fully later.

The Micronutrients: Biologically Active Non-Nutrients

Young (1996) describes micronutrients as a third generation of health foods. The first generation of products reflected the public's (or rather some of the public's) interest in healthy foods such as fruits and fruit juices, yogurts, and whole wheat breads as well as multigrain breads. This was followed by the second generation of food products best described as low-calorie, "light" foods with no or very little sugar or fat content. They were heavy on non-caloric sweeteners and fat substitutes or mimetics.

This third generation of health foods represents a distinct change of direction for nutritionists and consumers. Nutritionists know, and consumers are beginning to grasp, that some dietary constituents may be links to disease prevention. This understanding is as yet tentative and confused

components are in high concentrations. Dietitians and nutritionists are slowly developing a sound scientific body of data demonstrating mechanisms of disease prevention through diet.

Functional Foods: Prebiotics and Probiotics

Two terms deserve greater explanation because they have become prominent in any discussion of nutraceuticals:

1. Prebiotics are non-nutritive components in foods that have an activity which has a beneficial effect on an individual's health when consumed. Prebiotics are a hodgepodge of chemical entities that defy description and are found in fruits, vegetables, fish and animals, and herbs and spices. They are best simply referred to as the phytochemicals (except those found in animals).
2. Probiotics are defined by Salminen et al. (1999) as "viable microbial cultures that influence the health of the host by balancing the intestinal microflora and thus preventing and correcting the microbial dysfunctions." Probiotics are also foods (for example, yogurts) which contain live microorganisms, largely but not exclusively of the *Lactobacillus* spp., but also bifidobacteria (Hoover, 1993), enterococci, proprionibacteria, and some *Saccharomyces* spp. (Lee and Salminen, 1995; Knorr, 1998).

(Some authors make a clearer distinction between the prebiotics and phytochemicals. Phytochemicals differ from prebiotics both in their function in the diet and in the concentrations in which they are found in foods. Phytochemicals are found in much smaller concentrations in foods. This distinction will not be made here but readers should be aware of it.)

Current thinking suggests a strong complementary activity between prebiotics and probiotics. Prebiotics include the various dietary fibers and oligosaccharides. They are either acted upon by the gut microflora, or they stimulate the gut microflora positively in some manner, or these food components are necessary to maintain the gut microflora in a healthy and functioning state (Katz, 1999). By this stimulatory action these prebiotics allow the gut microflora, the probiotics, to overgrow or compete successfully with any pathogenic bacteria or harmful material entering with the food.

Table 5.1 lists some biologically active non-nutrients with their sources

Table 5.1 Some Biologically Active Non-Nutrient Factors Determined to or Believed to have Beneficial Effects against Some Disease Conditions when Consumed

Classification	Category and Food Sources
Probiotics	Bifidobacteria • Fermented milks, yogurt *Lactobacillus* species • Fermented milks (acidophilus milk); yogurts *Streptococcus* species • Fermented milks, yogurts
Prebiotics (phytochemicals)	Fatty acids • α-linoleic acid (canola and flaxseed oils) • Conjugated linoleic acid (safflower, sunflower, and soybean oils) • γ-linoleic acid (evening primrose oil) • ω-3 fatty acids (various unsaturated oils) Lecithins • Phospholipids (various oils, especially soybean oil), e.g., phosphatidyl serine, phosphatidyl choline Unsaponifiables of oils • Phytosterols (canola and soybean oils) • γ-oryzanol and ferulic acid (rice bran oil) Organosulfur compounds especially plants of the cruciferous (broccoli, cabbage, and cauliflower) and allium (garlic, onion, and leek) families • Isothiocyanates (mustard oils of Cruciferous vegetables) • Sulfides (e.g., diallyl disulfide) and oxides (allicin) from garlic and onions Terpenes • Monoterpenes: limonene, perillyl

Table 5.1 (Continued) Some Biologically Active Non-Nutrient Factors Determined to or Believed to have Beneficial Effects against Some Disease Conditions when Consumed

Classification	Category and Food Sources
	Polyphenols (including flavonoids and catechins)
	• Anthocyanins (blueberries, cranberries, tomatoes, red wine, tea, onions, kale)
	• Various other phenolic compounds
	Phytoestrogens (isoflavones)
	• Genistein, daidzein (soybeans, whole grains, berries, flaxseed, licorice)
	Fiber (and associated material)
	β-glucans (oats, barley, wheat, rice)
	• Lignans (flax)
	• Other soluble and insoluble fibers (fructooligosaccharides, e.g., inulin in Jerusalem artichoke)
	Saponins (derivatives of pentacyclic triterpenes)
	• Ginseng, soybeans, grains
	Herbal components
	• See, for example, Tyler, 1993; Anon., 1998d

Prebiotics

After the discovery of fiber's health-promoting properties, the search for other non-nutrients with health giving properties was on. Literally and figuratively the floodgates burst. The number of prebiotics with suspected health-enhancing properties grew rapidly and is still rapidly growing.

That chemicals found in plants have a significant effect on the human body should not have been a surprise. Plants have been used for ages by many cultures for medicinal purposes; this is a well-established observation. Drug and chemical companies have scoured ancient medical writings, old herbals, folk literature for herbal traditions, and native peo-

use them as raw materials to adapt as medicines, as food ingredients, or as initial building blocks for industrial materials.

Phytochemicals comprise a diffuse group including some micronutrients and a motley variety of chemicals with medical, stimulatory, flavoring, coloring, texturizing, antioxidant, non-caloric sweetening, narcotic, etc. properties. Some of the phytochemicals whetting the interests of food scientists and food manufacturers alike are (see also, Zind, 1998):

- Antioxidants. Vitamins C and E long known for their antioxidant properties have been used as such in foods. Both are reported to have a positive effect on the immune system and as well there have been reports of beneficial effects against cancer and cataracts (Elliott, 1999).
- Tocotrienols and γ-oryzanol. Both are sterols which are found in rice bran oil. γ-oryzanol is claimed to reduce plasma cholesterol, lower cholesterol absorption from the gut, and inhibit platelet aggregation (McCaskill and Zhang, 1999). The tocotrienols, similar to the tocopherols, are good antioxidants and also reduce cholesterol.
- Lycopene and carotene. The carotenoids are found in a variety of plant sources (Nguyen and Schwartz, 1999) but are especially plentiful in a common source, tomatoes. Carotenoids have been linked to a reduced incidence of cancer, particularly cancers of the prostate gland, lung, and stomach (Astorg, 1997). β-carotene also has antioxidant properties.
- Isoflavones. Genistein and daizein are proteins possessing estrogen-like properties which appear to inhibit cancer growth by stopping angiogenesis.
- ω-3 fatty acids which are found in the fats and oils of plant sources and also many animal sources (Ahmad, 1998). Some important fatty acids of this type and the related ω-6 type are palmitoleic, oleic, linoleic, and arachidonic. They play an important role in reducing the incidence of atherosclerosis and certain cancers.

Food manufacturers are eager to incorporate these into their food products and so promote the health benefits of their products.

Prebiotics are foods themselves or components of foods such as pea flour, sweet lupin meal, β-glucan, inulin, and other oligosaccharides especially those found in soybeans that have been used for centuries in many

Confounding this presumed value of the prebiotics in the diet, there is controversy over their effectiveness. For example, there is contradictory evidence from epidemiological studies that suggests fiber does not have the protective effect against colorectal cancer as claimed. Fuchs et al. (1999) in a study of 88,757 women between 34 and 59 years of age conducted over a 16-year period could not produce evidence of any protective effect of dietary fiber against colorectal cancers or adenomas. So even with the nutraceuticals there is confusion about their benefits in the diet.

Probiotics

The activity of added microorganisms, the probiotics, in the gut produces a beneficial effect on the health of the individual. The ways in which this benefit is derived are slowly beginning to be elucidated. The addition of probiotics to the diet can be done with fermented foods as the vehicle or with either live or dead cultures directly. Live microorganisms have been used in foods for some preservative action that they offer for many centuries and by many cultures (Fuller, 1994; Knorr, 1998), for example:

- Alteration or removal of a sensitive, unstable substrate in a food, e.g., desugaring of egg whites or winemaking
- Acidification of the food to pH ranges inimical to food spoilage microorganisms, fermented vegetables (sauerkraut, various types of kimchi, bean curds, tempe) and fermented meat products (various sausages)
- Production of some protective antimicrobial agents by the added microorganisms which when ingested provide protection against microorganisms of public health significance
- The addition of beneficial (and benign) microorganisms to overgrow potentially hostile microorganisms originating in the food

Yogurt and cultured milk products, without too much stretch of the imagination, fit all the above. Microorganisms are introduced as starter cultures in meat or dairy products or they are present naturally in a food and selected through salting procedures, i.e., sauerkraut manufacture or other fermentation processes. Kimchi has been touted as a functional food with both probiotic and prebiotic characteristics (Ryu and West, 2000).

Sanders (1999) reviews current thinking on the action and mechanisms of probiotics in the body. In addition, she lists strains of microorganisms with the reputed biological activity and diseases that they are believed to repress.

Brassart and Schiffrin (1997) discuss mechanisms for the beneficial action of probiotics and describe the possibility of designing products not only for healthy consumers but for clinical trials with sick people. They discuss both probiotic concepts of treatment using live micro-organisms and prebiotic concepts for treatment in which substrates (for example, oligosaccharides) on which the microorganisms feed are themselves fed to patients. Ishibashi and Shimamura (1993) describe the extensive applied research and product development that have occurred in Japan since the middle of the last century. Japan has been much quicker to adopt bifidus products (those utilizing Bifidobacteria) than have other countries.

Many areas of research for both prebiotic and probiotic components of foods remain to be explored further before effective new products can be developed for consumers in the new millennium to complement the enormous number of traditional fermented vegetable, meat, and dairy products that are available. Where, and at what site, in the gut are probiotics (and prebiotics) most effective? How can clinical trials using selected probiotics be undertaken (Brassart and Schiffrin, 1997)? Lee and Salminen (1995) discuss the need for more research to enhance the stability of probiotics and to define optimum concentrations for both traditional and new food products. These issues require answers before commercial advantage can be taken of prebiotics and probiotics.

Functional Foods: The Nutraceuticals

There is some dispute over the proper name for these factors. Many prefer that the term functional food be limited to foods containing prebiotics or probiotics, that is, be applied only to foods providing a health benefit. Nevertheless, nutraceuticals is the popular term in vogue; one used by many scientists themselves and is much more picturesquely descriptive of their value in nutrition.

In Table 5.2 some of the effects of pre- and probiotics are listed with proposed mechanisms of action. Because many prebiotics are strong antioxidants, it is believed that this property contributes to their biological activity. However, the simple and uninformative statement "boosts the

Table 5.2 Suggested Modes of Action of the Prebiotics and Probiotics

Probiotics	Prebiotics
• Overgrowth of pathogenic bacteria in the gut by benign bacteria • Detoxification of toxic or carcinogenic factors in the intestinal tract • Hydrolysis of lactose • Stimulation of immune system • Antihypertensive effect	• Provide factors for or stimulate growth of probiotics • Tumor suppressors • Act as antioxidants to remove toxic free radicals • Stimulation of the immune system • Block or delay progression of cell growth and precancerous lesions • Improve circulation • Have an antibiotic action against pathogenic microoganisms • Influence cognitive function • Reduce cholesterol

non-prescription medicines, food supplements, and herb-based ingredients to be used in or on foods as well as new products. There has been a frenzy of research activity to find and identify the active components, the phytochemicals in herbals and foods, that have the promised benefits. When these have been identified:

■ Isolated fractions containing the active components can be used as medicines.
■ Their presence in foods can be enhanced through conventional or unconventional breeding techniques, for example, carrots with more carotene (Zind, 1998).
■ Concentrates of isolated fractions can be used as ingredients to make finished foods.

At present, research is directed at developing methodologies for detecting and quantitatively analyzing their presence in foods. It is imperative that accurate methods be developed in order that:

Challenges for the New Nutrition

As Young (1996) has put it

"The science of nutrition has moved…to understanding the physiological and genetic mechanisms by which the diet and individual food components influence health and disease."

This new nutrition opens up a Pandora's box of ills and hopes. There are now opportunities to develop food products directed to specific health problems as, for example, anticarcinogenic foods or foods to improve cognitive ability or for use in particularly stressful situations such as occur in space flight, training for athletics, or in convalescence from trauma or surgery. There are already, for example, sports drinks, i.e., drinks designed to rehydrate the body rapidly or to provide energy or enhance energy metabolism during strenuous activity (Brouns and Kovacs, 1997).

But problems have arisen and will continue to arise for both food manufacturers and for legislators with these new opportunities.

Problems Presented by Functional Foods for Manufacturers

There has been a mad rush to capitalize on the consumer's quest for a long and healthy life free from debilitating diseases. It is obvious that interest in food factors that contribute to the prevention of disease, repair damaged DNA, have a hormonal regulatory effect, and enhance the body's immune system would have excited the scientific community and of food manufacturers.

The popularity of biologically active non-nutrients as subjects at scientific, technical, and trade conferences and exhibitions is immense. In the period from January 1998 to December 1999 there were more than 22 major events in 10 countries representing every continent except Africa and South America. These meetings were devoted to some aspect of nutraceuticals and functional foods as anticarcinogens, as marketing tools, or as food ingredients. They had titles such as:

Functional Foods: Beyond Vitamins and Minerals
Global Developments and Opportunities in Functional Foods and
    Nutraceuticals
Herbal Extractions as Food Ingredients, Medicines, and Supplements

professional food, trade, or international associations whose conferences may have had individual sessions on functional foods.

The motivations of both scientists and manufacturers, however, may be vastly different. Scientists want eventually to identify and isolate those phytochemicals in foods having the beneficial effects of providing a long and healthy life.

The individual manufacturer is caught in a difficult situation. On the one hand there is an opportunity to develop new products with unique health benefits. At the same time, unfortunately, these phytochemicals already have the attention of customers/consumers and competing food manufacturers. The health(y) foods bandwagon is a popular one that all manufacturers want to be on.

On the other hand, unfortunately for food manufacturers, many of the promises the new products offer have not been clearly established; their safety for all populations has not been determined; and new product development costs money.

Manufacturers want to be able to inform their customers/consumers of the benefits of their products. Yet labeling and advertising regulations must protect customers/consumers from the hyperbole that often accompanies advertising claims. Some examples of announcements, taken from the Institute of Food Technologists Web site in the early months of 2000, are

> "add more than value to your food and beverages…add life, with Polyphenols" to describe products put out by Templar Food Products
>
> "…herbally-active, protein enriched frozen juice bars" which contained chromium, manganese, 100% of the daily requirements of Vitamins A, C, and E, as well as protein. These products were from Cold Fusion Foods.

With the rise in availability of these products also came a warning from the health professionals of the dangers some nutraceuticals might pose for people, especially people on medications and young children eating nutraceutified candies and snacks.

Manufacturers face the problem of trying to inform the public of the benefits of a new product in a product area where, metaphorically, shifting sands prevent the establishment of a firm product identity.

Both scientists and manufacturers work, or should work, hand-in-glove.

Delivery Systems

Delivery systems of phytochemicals have taken many forms. One candy manufacturer was adding phytochemicals which were purported to prevent cancer, bolster the immune system, and reduce cholesterol in the candies it manufactured (Zind, 1998).

Ice cream has become a vehicle for supplementation. It has been flavored with green tea (a source of catechins), ginger, avocado, sesame, wasabi, and capsaicins (hot pepper principle) to become a nutraceutical. Fiber and calcium compounds have been added to beverages.

The lowly potato chip and other snack foods, such as corn puffs, have had herbals and plant extracts such as ginseng, St. John's wort, ginkgo biloba, or kavakava added to them (Abu-nasr, 1998). The promotional gimmick would be that they would promote long life (ginseng); improve memory (ginkgo biloba); combat depression (St John's wort), and aid relaxation (kavakava).

These products and others with added chemicals join a growing list of products with added calcium to enhance calcium intake, with added caffeine to combat drowsiness, with creatinine as an aid for body builders, with added fiber to assist in lowering cholesterol, and so on. The marketing gimmick is to provide old products with a new role by giving them a functionality. By providing soft drinks with calcium, protein, or vitamins, they have been moved from a refreshment role to a functional one as a nutrified beverage; snacks with either St. John's wort or kavakava are given a medicinal role.

Manufacturers need to convince customers/consumers of the benefits — and justify the extra cost — of the enriched products. That is, they will have to develop effective means to communicate the health benefits to a frequently skeptical and jaded public about medical claims or health properties of foods.

Problems for Food Legislators

For legislators, the problems will be twofold:

1. Safety. First and foremost, the safety of the specific phytochemical in the concentrations used in a food must be clearly established. Like any other additive, ingredient, or medication, safe limits for

Safety

Safety concerns are paramount. Phytochemicals are biologically active materials. Usage levels in prepared foods at concentrations higher than naturally found in a source food may not be safe for all populations. The presence of phytochemicals in common non-source foods, i.e., foods in which phytochemicals are not normally found but into which they have been added as an enhancement, may be harmful to some populations who eat a lot of the enhanced foods. For example, snack foods targeted for children or teenagers who are voracious snack food consumers may not be appropriate vehicles for supplementation with phytochemicals.

Chung et al. (1998), Kardinaal et al. (1997), and Hasler (1998) all have expressed a need for further research on the safety of functional foods in which the phytochemicals are found. Their concerns are many:

- How these phytochemicals work in the body is not clearly understood. What are the side effects?
- How nutrients in the rest of the diet might influence the activity of these phytochemicals or be influenced by them is unknown.
- There are no toxicity data on these substances. At best, there is only anecdotal information. Therefore there is no information on what acceptable dose levels might be for different segments of the population.
- There is also a danger of physiological interactions of phytochemicals with other medications that consumers might be taking at the same time.

Hasler, in particular, sees the need to balance benefits and risks with the use of foods containing physiologically active substances.

Chung et al. (1998), for example, review one phytochemical group, the tannins. These display rather ambiguous properties. On the negative side, tannins have hepatotoxic activity. They also display antinutritional activity since they can form complexes with digestive enzymes and proteins. The nutrient value of food is thereby reduced. Epidemiological studies have demonstrated high levels of mouth, throat, and oesophageal cancers in people using betel nuts, or the herb tea called *maté*, and who use sorghum as a staple in their diets. All these products are rich in tannins. Chung et al. also report on a study in which a positive relationship has been noted between the incidence of tea drinking and that of stomach,

138 ■ Food, Consumers, and the Food Industry

ambiguous effects of the tannins, Chung et al. (1998) suggested that the effectiveness of the tannins for health promotion might be dose-specific.

Such dose specificity presents legislators with a dilemma because there are good and bad sides. There is a confusing array of promises and denials that will accompany any research into food components with biological activity to be used as viable adjuncts to the diet. Much more research must be done to clarify the value, or the danger, not only of the tannins but of many other phytochemicals.

Herbs and herbal preparations present a slightly different situation. Here a different cautionary note must be sounded. These preparations do have a long, substantiated history in folklore as cures for some maladies and, indeed, herbs have been the basis for many modern medicines. Many of their active chemicals have been rigorously studied by scientists. They have through this history developed a veneer of respectability. Herbs are readily available in most health food stores as well as mainstream supermarkets where they are sold as preparations for teas or tisanes.

Tyler (1993) and others have suggested several dangers respecting the sale of herbs and herbal preparations:

■ Are they what they are claimed to be? That is, do labeled packages actually contain the herbs they are labeled with? Not all parts of herbs are effective; that is, roots, leaves, seeds, stems, flowers, bark, etc. may contain no or high concentrations of the phytochemical. Many countries do not have any regulations for the verity of herbal preparations.

■ Herbal preparations are not standardized with respect to the active ingredient(s) that they contain. Do they contain the active phytochemical in the concentration claimed and are they safe concentrations?

When such unstandardized products are used as ingredients in food products the question then arises: Are they safe for all segments of the population?

Tyler (1993) does not give all the herbals a clean bill of health. Ginseng has a very low level of risk even with excessive use. St John's Wort also is relatively safe but may with some people and at high dose levels cause a photosensitivity. Ginkgo biloba extracts do promote vasodilation with improved blood flow but very large doses can have unpleasant side effects.

At the present state of nutritional awareness possessed by the general

capitalize on "being the firstest with the mostest" with products to capture a market niche. How safe are these products for consumers and who is to judge the suitability of these products for consumers concerned about their health?

Scientists, who stand to benefit from the financial support of the food industry, and the food industry, which stands to benefit from the findings of scientists, need to find some means to communicate their findings and promote products based on these findings. Neither wants to disseminate misinformation to or endanger the health of customers and consumers.

Such knowledge as this new science of nutrition is bringing into the hands of entrepreneurially spirited hucksters and promoters has spawned a host of new food products. The U.S. government has relaxed restrictions on food manufacturers' health claims. Already warnings have appeared from the medical profession and scientists about the overuse of products containing nutraceuticals especially in conjunction with conventional drug therapy.

Regulation

It would be only natural that if a food substance had a proven, or even a highly suspected, role in combating a disease condition or in preventing its occurrence or was able to combat memory loss or depression, that there would be a strong motivation for the consuming public to want the food containing that substance. This need by customers/consumers would be sufficient cause for food manufacturers to produce attractive and tasty products with that particular component in them. And it would be equally desirable to want to tell customers and consumers alike that this beneficial substance was in the product.

Proper regulation can only come when safety data applicable for all populations of consumers and good analytical procedures are in place, and uniform standards for products have been developed.

Genetics and Nutrition

"Tell me what you eat, and I will tell you what you are."

A. Brillat-Savarin, *Physiologie du Goût* (1825)

140 ◾ Food, Consumers, and the Food Industry

states under all environmental conditions and all levels of activity. Earlier concerns were with the nutritional problems associated with undernutrition for which nutritional guidelines then in place worked reasonably well. They were not appropriate for nutritional problems associated with over-nutrition or the prevention or combating of chronic diseases. Thus guidelines had to be described for:

- Age of the consumer. Growing children obviously have more demanding needs as their bodies develop. Similarly, the nutritional needs of the elderly differ from those of young, vigorous adults and teenagers.
- Physiological condition of the body. Pregnant women, lactating mothers, and convalescing patients have more critical nutritional requirements.
- Geography and climate. People in tropical climates do not require the same caloric density of food as those living in arctic conditions.
- Activity levels. The heavy laborer, the athlete in training, the sedentary desk worker, the pressured, harried executive, the soldier on stressful military duty all have very different nutritional needs.
- Body composition. An individual's sex, body size, and weight also contribute to the nutritional needs of the individual.

These were all well-documented factors in establishing nutritional guidelines in the twentieth century. But even these were not enough.

Closer examination of feeding trials by nutritionists began to detect familial traits which led directly to the probability of genetic factors in nutrition. Why did heart disease run in families? Or overweight? Or high cholesterol? Or diabetes? Why, for example, would oat bran bring about a reduction in cholesterol in some people but not in others (Simopoulos, 1997)? Clearly, there was some other factor, perhaps a genetic contribution.

An interesting and statistically significant correlation was noted between the genetic taste marker, 6-n-propylthiouracil, and a preference for cruciferous vegetables and green and raw vegetables. People can be classified according to their ability to taste this marker: Some cannot taste it; some can taste it a little (medium tasters); and some react violently to the taste (strong tasters). Drewnowski et al. (2000) tested women with and without breast cancer. Those who were medium or strong tasters of the genetic marker disliked the vegetables most. They were less likely to have eaten

foods declined as her taste perception of 6-*n*-propylthiouracil increased. No similar observation was noted in men.

Phenylketonuria has been known for many years. This is a genetic disorder in which those afflicted cannot assimilate the amino acid phenylalanine. Cystic fibrosis is also a genetic disorder. It has long been recognized that individuals with cystic fibrosis are fat malabsorbers. Now research conducted by Alvarez and Freedman on mice genetically altered to have cystic fibrosis has shown that the fatty acid docosahexaenoic acid reversed these symptoms in mice (Maugh II, 1999; see also, the Cystic Fibrosis Foundation's Web site http://www.cff.org/news9910a.htm). This was reported at the North American Cystic Fibrosis Conference in Seattle on October 9th, 1999.

The past 20 or so years has seen a growing awareness of genetic variation and dietary interaction in the treatment of many chronic diseases. The result has been a marriage of the sciences of medicine, nutrition, and genetics with food science to develop pleasing and acceptable quality products to meet dietary needs. Some of the conditions listed by Simopoulos are detailed in Table 5.3.

Table 5.3 Chronic Conditions with Genetic and/or Nutritional Linkage

Chronic Condition
• Coronary arterial diseases
• Serum cholesterol levels
• Obesity: affected by hunger mechanisms, satiety, or body metabolism
• Diabetes
• Blood pressure
• Lactose intolerance
• Some cancers (breast cancer) with familial traits

An individual's genetic make-up can be a determinant (Simopoulos, 1997):

- ■ In nutrient absorption
- ■ In metabolism and the excretion of metabolic byproducts
- ■ In taste perception and hence food preferences
- ■ In the degree of satiation

environmental risk factors), then there is a clearer picture of how diet and genetic make-up can influence a person's health.

Obesity can have a genetic basis but as Foreyt and Poston (1997) point out, obesity is "… a multifactorial disorder with multiple causes." They report that at least 20 genes might be factors in some types of obesity. Environmental factors (for example, level of activity) do play a strong role in obesity as well.

In the third millennium consumers, from childhood to mature adulthood, will be assessed for what diet is best for them based on their genetic make-up. The technology will be available to prescribe diets for individuals based on foods containing all the correct factors to reduce the risk of diseases that their familial history and genetic make-up indicate they might be subject to (Patterson et al. 1999). By understanding the molecular mechanisms in health and disease, diseases can be subdivided more specifically and treated accordingly with a more directed dietary intervention. Shades of George Orwell's Big Brother watching one!!! Instead of the Thought Police described in Orwell's *1984* it just might be the Diet Police who are watching.

Incorporating the findings of genetics into nutritional guidelines for consumers and, at the same time, developing nutritionally suitable, quality food products that are appealing in taste and texture will present major technical, communication, and logistical problems. It creates, as Dodd (1997) puts it, the personalization of dietary guidance.

Medical Foods

The previous topic leads directly to the subject of medical foods. These are designer foods of a unique nature. They are nutritionally controlled foods intended for hospitalized patients suffering severe trauma who must be fed either intravenously or by a nasogastric tube or for individuals who for a variety of reasons may require special diets. Some reasons described by Schmidl and Labuza (1992) for the need for medical foods are presented in Table 5.4.

Schmidl and Labuza (1992) classify these foods into four distinct groupings:

1. Nutritionally complete products which provide all the required

Table 5.4 Conditions for which Nutritionally Designed Foods may be Required

Conditions Requiring Nutritionally Designed Foods
• Severe trauma resulting from physical injury, cancer irradiation therapy, burns, etc.
• Malabsorption of nutrients resulting from intestinal resection or disease
• Gastrointestinal organ malfunctions
• Severe allergies requiring avoidance of or very controlled amounts of nutrients
• Genetic errors of metabolism
• Physical state of the individual, i.e., immobility

3. Foods for metabolic disorders which are designed to provide the nutrients required while minimizing the adverse effects of the nutrients responsible for the disorder
4. Oral rehydration solutions which are designed to replace water and electrolytes lost in some diseases

The actual composition of these diets will vary widely with the particular disabilities of the individual. Schmidl and Labuza provide an extensive listing of some ingredients used in their preparation.

Salminen et al. (1999) give information on the use of probiotics to alleviate the symptoms of lactose intolerance and some diarrheas. Duran et al. (1999) report on the importance of the management of blood phenylalanine levels with a phenylalanine-free amino acid mixture. Their sample comprised patients hospitalized because of careless personal home treatment.

As organ transplant operations become more common, there will inevitably be nutritional complications for recipients, especially those receiving liver, kidney, and intestinal transplants. Silver and Castellanos (2000) discuss a case study involving the nutritional management of an intestinal transplant in a child.

The last two references introduce interesting challenges. Organ recipients as well as those with genetic metabolic disorders will require medical foods with uniquely designed nutritional requirements but which must also be appealing taste-wise, visually, texturally, and are safe from a

create such desirable products for victims of genetic and chronic diseases, or suffering radical surgery to control or alleviate their infirmities.

The Quest for A Long and Healthy Life

There is a natural desire on the part of the general public to want to live a long and healthy life free of chronic diseases. Some consumers will try, within certain limits, to adopt foods or activities that promote such a healthy life. The vicissitudes to one's length of life and to one's general well-being brought on, for example, by obesity or overweight, have brought out a number of countermeasures which many have adopted with more or less success:

- Physical activity. By becoming more active people can hope to keep their weight down and their physical well-being improved.
- Better nutrition through better diets. Adherence to recommended dietary guidelines plus information from low fat cookbooks and books describing better health through special diets all prompt people to watch their diets. Restaurants now promote low fat dishes or vegetarian items on their menus.
- Lifestyle changes. Attempts have been made to change lifestyles Thus was born the Western embracing of ancient Asian and oriental philosophies. These were not new philosophies but were new to most in the Western world. Their practice brings calmness and spirituality to the individual and is an aid to meditation.

Unfortunately many of the diets that are promoted in the diet cookbooks are written by non-professionals or by authors with questionable motives in writing them. They have some gimmick or personal theories they wish to promote. These cookbooks are best described as containing fad diets.

Physical activity became a panacea for many ailments afflicting the body. Running, tennis, swimming, hiking, aerobics classes, gymnastic dancing, stationary bicycles, stepping stairs, and treadmills all became vehicles to burn calories, condition the heart, lower cholesterol, and combat the onset of osteoporosis. All this was in the name of a healthy and productive life well into old age.

Prudent menu choices, items having low fat or low salt or no

as being as important in good nutrition as are those fats rich in ω-3 fatty acids.

Changing lifestyles of individuals is difficult. Outside intervention as well as the personal commitment of the individual are required to manage such a change. Nonetheless, people, some 30 years ago, were answering questionnaires to determine if they were type A or type B personalities. Type A personalities were characteristic of those individuals who were aggressive and driven, in effect, workaholics, and were thought to be potential victims for a heart attack because they could not control their stress and anxiety levels. Type B traits were marked by more complacent, calm, serene and relaxed attitudes. People with these Type B traits were considered less likely to have heart attacks.

Toops (1999) extravagantly headlined an item describing philosophical movements as food spiritualism and in support of her assertion described a series of Zen-themed cookbooks. Classes, associations (variously described as spiritual wellness centers), even political parties (the Natural Law party, for one) associated with these philosophies have sprung up. As adherents to these philosophies grow in number, food manufacturers should look carefully to take advantage of these trends not only with food products but with the paraphernalia associated with their practice.

For those embracing these philosophies to change their lifestyles and to improve their health, there were myriad secondary and even tertiary demands in addition to meditation. Many of these philosophies are also associated with certain ritual activities and exercises such as yoga, tai chi, etc. The practice of these exercises and teachings was considered to aid the individual in controlling stress, lowering the blood pressure, and helping the individual to live longer. Particular foods and diets associated with the cultures that embraced these philosophies, faiths, and religions were recognized and popularized by food manufacturers and restaurants.

In some small measure these philosophies, plus the general concern for less fat in the diet, have spurred an interest in vegetarian cuisine or certainly in a cuisine less dependent on animal protein as a major component in the diet (Wrick et al., 1993). (In developing countries, the interest is in adding some animal protein to their diets.) This, in its turn, has led to opportunities for food manufacturers for the development, promotion, and sales of new foods and new food products adopted from these philosophies as well as the publication of diet and recipe cookbooks based on them and adapted to Western society's culinary habits.

Exploiting the Consumer's Quest

The exciting possibilities for functional foods in the control, amelioration, or slowing down of chronic and degenerative diseases will certainly pit two unlikely foes: food companies and pharmaceutical companies. The two will approach the issue from very different promotional aspects. For each it is a new venture. The pharmaceutical companies will venture into healthy, good-for-you foods while the food companies venture into medical foods, i.e., into patent medicine foods.

There are several terms to describe these new products: functional foods, designer foods, and medical foods, and they can take several forms:

- Dry for making teas and tisanes
- Beverages such as prepared flavored teas, tisanes, soy milks, nutriceutified juices
- Dairy products such as yogurts and fermented milks
- Snacks flavored with herbal preparations
- Supplements in the form of capsules, tablets, and liquids (patent medicines)
- Main dishes (soybean veggie burgers)

Wrick et al. 1993 conducted a telephone survey that showed older respondents were more aware of the health benefits of functional foods containing phytochemicals than were younger respondents. This older group was also more likely to believe in the efficacy of these products. Wrick et al.'s study also indicated that the preferred way to deliver these phytochemicals was by increasing the consumption of fruit, vegetable, and cereal products followed by the corollary suggestion to increase the content of these foods with phytochemicals by developing special varieties.

The least desired ways of delivering the beneficial ingredients were with pills or capsules, fortified snacks (cookies, candies and beverages), or additions to the water supply.

Some of the concerns that food manufacturers have about commercially exploiting these foods are discussed in depth by Kuhn (1998). These have been touched upon already:

- Safety of functional foods and phytochemicals
- Extracts or concentrates vs. the natural food
- The ethics, and safety of, fortifying snack foods targeted for children

months last year, almost 1,000 contradictory food and nutrition articles were reported." Such a statement staggers the imagination. Many public and high schools have removed nutrition from their curricula. Governments have removed financial support for health and physical activities and gone, too, are many school milk programs.

Nutritional knowledge is largely dispersed to the general public through various channels: cooking columns in newspapers, nutrition and diet columns in magazines, radio and television food shows, and now the Internet. It is here that the public learns that cardiovascular diseases, heart disease in particular, high blood pressure, certain cancers, diabetes, and several other diseases may be caused by poor nutrition. They learn that heavy consumption of alcohol can lead to cancer of the esophagus and of the liver and that a high fat diet can be implicated in breast, pancreatic, and some intestinal cancers. They also learn that one or two alcoholic drinks a day are good for one's health; red wine prevents macular degeneration, etc.

Consumers are also told that the advent of these disorders may be prevented, deterred, delayed, or somehow the odds of getting them may be reduced by good nutrition or even, and here is the rub, by the consumption of specific foods. For example, fiber from various sources has been described as beneficial for lowering cholesterol in the blood and as a preventive against some intestinal cancers. Vitamins A, C, and E are reported to reduce the incidence of some cancers. Indeed, vitamin E, as reviewed by Ahmad (1996) has a protective role:

- Against free radical formation in exercise and in the aging process
- In the development of heart disease
- In the maintenance of structure and function in the nervous system
- In cataract development
- In the proper functioning of the immune system
- In maintenance of the skin against the deleterious action of the sun
- In cancer prevention

Soybeans and products derived from them (genistein, for example) have been reported to reduce the risk of certain cancers.

That is what customers and consumers alike are being told. It is unfortunate that information, misinformation, and confusing information about nutrition can be also disseminated to the public by these very same media. A practical example of the information available to consumers about

148 ■ Food, Consumers, and the Food Industry

First, there was a correct, but perhaps misleading, headline — misleading because of its brevity: "Herbs linked to fertility risk." The study was done on hamster sperm and eggs.

Then there followed the subhead: "Studies show high doses of herbs, including the popular echinacea, have an ill-effect on fertility in hamsters. But scientists say they don't know the effects on humans." Correct so far, but again there is no mention that this was an in vitro study.

The study (Ondrizek et al., 1999) was referenced in the article by Owens by journal and lead author. (In addition, the lead author was interviewed by Owens.) Ondrizek's team found that high doses of three herbs, echinacea, ginkgo biloba, and St. John's wort damaged reproductive cells and stopped sperm from fertilizing eggs of hamsters in in vitro studies. This later finding was duly reported in the article. Owens also reported in the article that Ondrizek admitted that the effect of these herbals on human fertility was unknown. A further disclosure was that St. John's wort resulted in DNA mutations. The leader of the team was reported to say that more studies, including human studies, were necessary.

(Parenthetically it should be reported that

■ Echinacea has a well-substantiated reputation for its immune-stimulant properties.
■ Ginkgo biloba extracts have a well-documented history of usefulness in treating conditions caused by a decreased cerebral blood flow (usually touted for help with short-term memory loss).
■ St. John's wort has gained a respected reputation for usefulness in treating anxiety and depression.

The foregoing information for these herbs is from Tyler (1993), a respected and recognized authority on plant drugs.)

Manufacturers of herbal preparations were reported (by Owens) to cry "unfair". Owens interviewed one executive of the largest manufacturer of herbal preparations in Canada who stated that there was no cause for public concern but rather a need for more research studies. He claimed there was no evidence the ingredients in these herbal preparations ever reached the reproductive system (when taken orally).

understood the implications of the entire article? In particular, how will consumers react to snack products to which manufacturers are adding some herbal preparations?

The Internet, an increasing popular source of information, can, unfortunately, be both a rich source of reliable information and a repository of false and misleading information about nutrition. An Internet search for the phytochemical, oryzanol, provided many interesting and somewhat outrageous claims for the product's efficacy from suppliers.

Who is Responsible for "Responsible" Nutrition Education?

This is a question that seems to have fallen outside everyone's area of responsibility. Governments, i.e., those responsible for public education, seem to have no interest in subjects like nutrition, health, and physical activity, or at least, pay only lip service to these topics. As money for education gets tighter, these are the items that suffer most by deletion from the curriculum.

Governments do establish nutritional guidelines and attempt to educate with pamphlets and brochures. These are distributed throughout the school system.

Childs, reported by Busetti (1995), describes the news media, then the medical community, followed by government nutrition awareness campaigns as the major vehicles for the consumers' awareness of nutrition. Manufacturers' advertising was the poorest vehicle for conveying food health benefits to consumers.

The validity of these vehicles for education is highly variable. The media thrive on controversy and the "breaking story." The story must be told in 10-second sound bites. For accuracy in educating, television and radio are not good (the so-called infomercials now so common on television should be classed as manufacturer's advertising and not considered authoritative). Print media can do a more in-depth treatment but it is doubtful how well in-depth stories are understood by the general public despite competent reporting (see previous section). Again, reference must be made to Toops' statement about the frequency of contradictory nutritional stories (Toops, 2000).

The medical community and government resources then are the only reliable sources for nutritional information that the average consumer has. Thompson (Count Rumfound) was astonished at the neglect given to

Chapter 6

The Challenge of Ever Newer Technologies

"We must beware of needless innovations, especially when guided by logic."

Sir Winston Churchill

"Every step by which men add to their knowledge and skills is a step also by which they can control other men."

Max Lerner

"Everywhere we remain unfree and chained to technology, whether we passionately affirm or deny."

Martin Heidigger, *The Question Concerning Technology*

An Accelerating Pace of Technological Change

Tables 1.2 and 1.3 clearly demonstrate the rapid growth in science and technology that has occurred in the past 200 to 300 years — an almost exponential growth that has touched all industries, all governments, and

the past century. Technology is being touted as the answer to all the world's problems — "better living through science" has become the catchphrase of the day. Some futurists have even predicted that in the third millennium many government systems might become elaborate computer-controlled technocracies whose governance and economies will be directed by the application of elaborate data mining technology coupled with real-time economic data fed into their systems for judgement making.

Technology and science build on the experiences and observations remembered or documented by hunters, fishers, farmers, priests, shamans, medicine men, that is, by specialists. With each succeeding generation the substance of these observations becomes more refined with new observations or is demolished as previously accepted beliefs prove false. New theories emerge and the next generation of scientists/observers goes off on tangents adding their experiences and observations to create new bodies of accepted beliefs. Students and apprentices of this new body of beliefs apply their knowledge to fit their daily requirements for surviving and enjoying life. Gradually, these refinements in knowledge develop into the various bodies of science known today. Each moves forward in fits and starts and often disconnectedly with the advances made in the sciences and the structured investigations that support these sciences.

Accum (1769–1838) was a brilliant chemist with an interest in many fields but especially in food chemistry. He became a thorn in the side of English food manufacturers, especially the brewers, as he reported on the adulteration of the English food supply in 1820 (Farrer, 1996; Farrer, 1998). His studies in food analysis began the long process resulting ultimately in pure food regulation for the food industry. There followed Justus von Liebig (1803–1873), who is generally acknowledged as the father of food chemistry, and who published his major works from 1837 to 1855. Lind (1716–1764) began the correlation of food to nutritional diseases. Appert's work on preserving "...All Kinds of Animal and Vegetable Substances..." appeared in the early 1800s and here began the art of thermally processing foods.

This period, spanning barely more than 200 years, represents the crystallization of several diverse sciences applied to food and agriculture. The science of foods is comparatively new as a separate discipline. Indeed, many of today's food scientists, especially the older ones, are graduates from other disciplines, e.g., biology (botany and zoology), physics, and chemistry. Fledgling food companies needed a greater

The New Science

The food microcosm has lagged behind somewhat in applying the developments arising in this burst of scientific growth in the past 200 years. Certainly developments in other technical fields had both indirect and direct applications to agriculture and fisheries, upon the food industry and its products, upon the retailing of food, upon the conductance of the food business and its management and, finally, upon customers and consumers. Nevertheless, the food microcosm was a few years behind in applying such advances.

New technologies and the emergence of inventions and processes have always been plagued with three stages of development:

1. Initially, there is resistance to adopting any innovative technology. Vested interest groups, often those practitioners of established technologies with which the new inventions conflict, can be counted upon to resist applying any innovative technology.
2. This initial resistance is usually followed by a second phase. Intrepid entrepreneurial users of the new technology develop secondary applications in which the innovation finds its own application niche. For example, the microwave oven was touted for the processing of foods and particularly to replace the kitchen oven. Its greatest use ended up as a convenient tool for the reheating of coffee, the serving of frozen entrees (the development of which the microwave oven gave impetus), and the reheating of leftovers from previously prepared meals. It has not had great success as an oven.
3. In the third phase, one finds that the innovative technologies become additive; new technologies are added to the older technologies. Old technologies are not replaced; they are improved upon, modernized, and refined. Freezing technology was vaunted to replace thermal processing because of the improved quality and nutrition of food that were obtained. It did not, but frozen foods did become another option alongside canned foods for customers. Flexible pouch packaging technology would see the pouch as a replacement for the rigid tin can. It, too, did not replace the tin can but merely became another packaging option for customers.

New technologies become additive technologies finding new niches

Kentucky was destroyed in 1849; it did damage to the air and to the crops, and its presence was blamed for the lack of rain at that time (Forbes and Dijksterhuis, 1963). With hindsight this sort of opposition appears ludicrous today. It is, however, not very different from the man-like apes or ape-like men caricatures that were used to deride Darwin's work on species or the opposition to irradiation in food preservation for fear of "glowing in the dark" as expressed by one food columnist.

The General Impact of Science and Applied Technologies

That technological developments can have effects far beyond their immediate impact was seen in Chapter 1. Advances in computer technology are a good example, having applications to all industries. The concomitant development of sophisticated software plus advances in communications technology have permitted the conduct of business at a much faster pace. Real-time data observance and real-time decision making have arrived.

The Computer, Computer Applications, and Communication Technologies

The impact of new computer applications was slower to be felt in food processing than in other industries. In these non-food-related fields, people saw how these applications could lead to better and more economical ways of doing ever more complex tasks; for example, communicating electronically for business activities such as order-taking, just-in-time distribution systems, inventory control, and financial transactions. It is now estimated that 70 to 80% of all financial transactions are conducted electronically; that is, no money is actually transferred.

In some sectors of the food microcosm, e-tailing, for example, as has already been described, there has been no reluctance to adopt new computer and communication technologies (Table 6.1). Computer-assisted communication between producers, manufacturers, retailers, and their customers has assisted in the development of a more rapidly responding, efficiently operating food microcosm, for example:

■ Faster access to technical information by agronomists, food manufacturers, and warehousing/distribution centers for problem

- More economical research and development testing to reduce the time for getting new products to customers
- More efficient tracking of sales to keep popular and rapidly moving items in stock and to assure proper inventories are available for promotions, to maintain better control of inventories to lower the monies tied up in inventories, and to monitor accounts receivable and other finances.

Computers have now become indispensable in the conduct of all business.

Food Applications: The Tortoise Meets the Hare

That section of the food microcosm roughly described as the food manufacturing plant floor has generally been much slower to adopt new technologies for the production, processing, and preservation of foods than has other sectors within the food industry. It is not an aversion to new technologies that restrains acceptance of novelty by food manufacturers. A perceived value or advantage for the adoption of a new technology must be clearly demonstrated before novelty can be adopted.

To be accepted any new processing technology must satisfy several criteria as outlined in Table 6.2. Some examples of food processing developments and what happened to them illustrate the need to satisfy these requirements:

High-pressure processing for preserving milk was demonstrated a century ago (Hite, reported in Hoover et al., 1989) yet it has only comparatively recently been used commercially, especially in Japan, where it has been used for fruit products. At the time of Hite's demonstration of this technology, there was no perceived consumer-driven need nor was there any marketing pressure being exerted by any company to establish a new market for high-pressure-preserved foods. An ancillary factor for delay in accepting the new technology was that there was no need to develop high-pressure processing equipment.

When the new technology is so novel that there is no off-the-shelf equipment that can be readily purchased and quickly applied, then the cost for custom-built equipment can be staggering for companies wishing to adopt the technology. The lack of availability of equipment with a reliable work history at a reasonable price deters many companies from investing in new and unproven technologies. There was, consequently, a

Table 6.1 Examples of Computer-Based Technology and Communications Uses in the Food Industry

Application	Execution of Application
Information retrieval	• Weather information for farmers • Veterinary and animal nutrition information • Accessing information bases for technical, legal, and nutritional information for, e.g., labeling purposes • Use of data-mining techniques to determine trends in databases • Creating artificial intelligence databases derived from employee knowledge and experience
Manufacturing	• Process assistance in, e.g., least cost formulation, optimization of product formulations, evolutionary operations • Analysis of process and quality control data in real time • Computer-controlled process operations
Marketing, sales, and distribution	• Analysis of consumer research and survey data; cataloguing of consumer communications about product questions, applications, and recipes • Analysis and management of consumer complaints for product and process improvement and weak points in distribution system • Inventory control • Design of product labels and packages • Order and shipping scheduling for JIT (just-in-time) delivery to retailers
Research and development	• Modeling procedures for process design and optimization • Modeling procedures for shelf-life stability studies

Table 6.2   Criteria to be Met by Food Manufacturers before the Adoption of New Technologies

Origin of Criteria	Criteria Required before Adoption of New Technologies
Customer/consumer driven	• Need for adoption must be consumer driven • Adoption of technology must offer some obvious advantage to or be seen as satisfying some customer/consumer need • New technology must be safe or provide greater safety, stability, quality, and nutritive value • New technology must itself be ethically, morally, socially, ecologically, and environmentally acceptable
Manufacturer driven	• New technology must provide an economic advantage measured by greater productivity, reduced waste, or greater efficiencies • Adoption must confer a marketing and/or sales advantage • Technology must be proven and be commercially feasible (equipment for applying the technology should be available as a shelf item or technology should be simply applied) • Funds at a reasonable cost must be available to apply the technology • Must be permitted by legislation at local, regional, and national levels
Marketplace driven	• Adoption provides a competitive edge • Competitors have already adopted or are considering adopting the new technology

Thermal processing technology was introduced into a vastly different

(anyone who could preserve foods). No new equipment was necessary. Napoleon needed an assured supply of stable food for his armies.

Other countries, fearful of Napoleon's military intentions and his desire for ruling Europe, were faced with the same problem of military feeding. Armies could no longer live off the land as they had historically. They quickly adapted Appert's technology for victualling their armies and their navies. Their plans for expeditions of discovery and territorial aggrandizement in the New World were greatly assisted by the stores of food they could carry on board their ships. Money and financing were not a problem since government funding was readily available and private entrepreneurs were motivated by profit. In addition, ancillary technologies (development of the steam engine and tin can making) were becoming available to complement the new technology with the development of the autoclave.

The application of irradiation technology to food processing is a classic example of antipathy toward a new technology. The reasons for this aversion and slow acceptance are identified in Table 6.2.

- Despite the potential benefits of irradiation for prolonging the shelf stability of some foods, it is not a socially acceptable food preservation technique for many customers and consumers.
- Many customers and consumers cannot identify, oddly enough, a critical consumer-oriented need for the technology. What they see are benefits of irradiation for food producers and manufacturers but not for themselves, the customer/consumers. The argument is "why don't they (e.g., poultry processors) clean up their processing operations" and "why do we get to have older, staler, and less fresh food because they want to nuke it for longer life?"
- This unacceptability to customers/consumers exacerbates the problem for the general public for the siting of irradiation plants. Simply put, no one wants them in their backyard.

Irradiation facilities are cursed with high installation costs because of the need to shield both workers and the environment from radiation leakage. They are, therefore, usually stand-alone plants leased by their owners. Other food companies must transport their products for irradiation and then back to their warehouses or distribution centers. This is costly. For all these reasons, there has been a long delay in the acceptance of irradiation since the first discovery of its potential for food preservation

Without any preamble, he quipped that one should avoid irradiated fruits or vegetables unless one "wants to glow in the dark!!"

Neither the food industry as a body nor food industries individually has made any concerted effort to educate the public to the advantages and harmlessness of food irradiation. Whether this reluctance is from fear of a public backlash or merely just a desire to keep one's head in the sand is a moot point.

To get the acceptance of newer food technologies, e.g., genetically modified foods, the food industry will have to come out of its collective shell and educate its customers. In addition to the customer/consumer, the industry must make new technologies appear natural and acceptable and not in conflict with the social and traditional mores of society.

There is another *caveat*. Any new technology, to be acceptable, must be necessary or be seen as necessary by the public. Churchill's words which began this chapter bear repeating:

> *"We must beware of needless innovations…"*

Both technologies, irradiation and the use of genetically modified foods, fall short, in the public's opinion, on this count. For example, opponents to irradiation to reduce the incidence of *Salmonella* microorganisms feel that the poultry industry should first clean up the contamination in the poultry feed supply before resorting to irradiation (see, for example, Coghlan, 1998b). Opponents fear, and anti-irradiation groups promote this fear, that application of irradiation in poultry processing will merely allow that industry's continuing acceptance of practices that lead to contaminated poultry products.

Neither has the case for necessity been made to justify genetically modified foods. For instance, plants modified to resist insect pests, to withstand herbicides and/or pesticides, or to withstand adverse growing conditions (i.e., to be able to be grown where it is presently too hot, too dry, or too cold) are basically grower-benefit crops and chemical company-benefit crops. For example, in the 10-year period ending June 30, 1997, 78% of USDA permits and notifications for genetically engineered plant field releases were for grower-benefit characteristics (Wilkinson, 1997) such as insect, viral, nematode, etc. resistance.

When the biotechnology/seed industry publicized widely that it was going to develop crops with a built-in sterility gene, the so-called terminator gene, genetic engineering was widely seen by both the general

forever dependent on the biotechnology/seed companies for their crop seed. Promises that no such development would be undertaken were quickly announced by these companies after the public outcry that arose.

Control over genetically modified seed as effective as the terminator gene would have been is the requirement that farmers could not carry seed over from previous year's crops for replanting in subsequent years. Thus farmers were tied in contractually to buying new seed each year.

Opponents to genetic engineering of crops fear biotechnology companies have gained too much control of farming practices through their dominance of the seed market, by their headlong rush to purchase seed companies, and through their ownership of the patents on genetically modified seeds. Many also fear a domination of monoculture practices by these companies.

Adoption of new technologies by the food industry and their suppliers for the production of foods and food products must be done rationally, ethically, sensibly, and needfully or, as stated earlier, be seen to be done thus before these will be accepted by customers and consumers. How these criteria for acceptance will be rationalized in the future remains to be seen but it is an issue the food industry collectively or food companies individually must face.

Old Technologies; New Hazards

The global village is a concept that has become a reality in one sense. There is an ease of movement. People move about more as tourists, as migrants forced to move, and for business purposes. Ideas are exchanged as are foods, food traditions, and food tastes. Tastes for foreign foods develop and foods from exotic countries are distributed worldwide.

It is inevitable in all these exchanges that some unwanted development must occur.

New and Improved Health Hazards

The Harvard Working Group on New and Resurgent Diseases (Levins et al., 1994) describes the emergence of new diseases. These are diseases which gain rapid prominence due to various socioeconomic factors (rapid urbanization, for one) or the creation of environmental changes which in

Desowitz (1987) describes several instances where humans have interfered with the environment, usually with good intentions, only to create a new disease vector with disastrous consequences to local peoples.

Not all the diseases identified by criteria established by the Harvard Group are food related. The criteria used have a general applicability to unusual or unexpected food disease outbreaks that could develop as new (i.e., lacking a long-established history) preservative technologies are applied singly or in concert with traditional technologies to new classes of foods. Some of the possibilities they list are

- As life spans increase in the population, a slowly evolving disease is able to develop and will become recognised only in its later or critical and devastating stages, e.g., variant Creutzfeldt-Jakob disease and some cancers.
- A rare disease becomes common (through, for example, rapid travel of tourists) or a disease confined to one locale or one population begins to spread rapidly as populations move.
- The virulence of what was a mild disease increases or the disease develops new characteristics.
- Improvements in diagnostic techniques allow the discrimination of new pathogens, e.g., prions, or an indeterminate malaise becomes more defined with recognized vectors.
- New populations or social groups are examined by medical professionals.

Some of the new diseases the Harvard Group lists are water-borne and therefore might have direct applicability to some imported food products (Levins et al., 1994).

There are an increasing number of food-related disease outbreaks that do meet some of the criteria that the Harvard Group established, for example:

- Alfalfa and bean sprouts (introduction of a comparatively new food) resulted in salmonella infections in February 1996. There have been outbreaks in Sweden, Finland, Denmark, and the U.S.
- *E. coli* 0157 (increased virulence of a known microorganism) was the causative agent in the deaths of four children who had eaten undercooked hamburger meat at a fast food restaurant in the U.S.

All these outbreaks were in countries with well-developed food infrastructures. Where, then, is the problem?

Old Abuses: New or Old Hazards?

The following illustrates a concern respecting hazards in the global village.

For several years, customers, particularly in the developed countries, have tended to want minimally processed foods. These are foods processed to be as close to being fresh as possible and which are available in all seasons. These foods are normally only processed by chilling to reduce field or body heat, trimming or cutting, sizing, washing (often in chlorinated water with water rinses afterward), and packaging to prevent recontamination or gas packaging with suitable mixtures of gases that inhibit the growth of spoilage microorganisms. Fresh meats receive a similar treatment. They are usually vacuum packed or packed in a carbon dioxide atmosphere. Some specialty dishes may have a mild heat treatment and have marinades or lemon juice added which confer some preservative action.

These minimally processed foods depend on properly chilled handling throughout the distribution chain. Any break in this multicomponent preservative chain (dubbed hurdle technology by Leistner and Rödel, 1976) will result in microbial spoilage or the development of other microbial hazards with public health significance. The system cannot withstand any abuses.

One abuse to the system identified by the Harvard Group is also listed in the food outbreaks above. It is the global village. With the advanced identification technology available it was possible to type the shigellosis microorganism responsible for the outbreaks recorded above that struck in Ottawa, Calgary, Cranbrook, B.C., Boston, Los Angeles, Miami, and in Minnesota. The contaminated parsley was identified through the shigellosis microorganism on it to have come from a farm in Baja California. It had been washed and cooled in contaminated tap water (Heinrich, 1999). Food, in the future, will come increasingly from areas around the world that might harbor food contaminants prevalent in those countries. To have fresh produce on a year-round basis in northern climes requires that it be imported from southern climes where agronomic practices and processing technology may not meet standards of other more developed countries.

at noon and later for the evening suppers without further chilling. Needless to say, this abusive treatment permitted the microorganisms on the produce to flourish.

The preservative chain in the production, distribution, and preparation of this parsley was thus broken at two vital links: in pre-processing at the source country and in the final presentation. Trained food handlers at the food service outlets perhaps could have limited the extent of the outbreaks. It is here, in the education of its employees, that the food service industry must in future accept greater responsibility.

Brody (1998) discussed at length the kinds of abuses that can occur in preservative systems used for the customers' growing demand for minimally processed foods which are available year-round. He described one paradox whereby modified atmosphere packaging, a technological advance, in association with abuses to the system can increase microbiological hazards (see also, pages 82 and 106 in Fuller, 1994):

> *Prior to modified atmosphere packaging, a customer could clearly see whether the product being selected was or was not spoiled. Mould growth or sliminess was obvious. The overgrowth of these overt spoilage microorganisms also had served as a deterrent to the growth of microorganisms of public health significance. With the advent of modified atmosphere packaging, microorganisms responsible for this obvious sign of spoilage were held in check. Now spoilage was retarded: the microorganisms of more serious health concern were left unchecked and were invisible.*

With virtually every corner of the whole world able to serve as a bread basket to local markets or to larger sections of the country or to the rest of the world, there is the possibility of food epidemics on a scale never seen before. And these will become more severe and more frequent. Rapid transportation of produce from all over the world plus the great increase in international travel for business and tourism will disseminate new food contaminants and food-borne diseases farther and faster than ever before.

New Technologies: What Hazards?

The major new technologies that will dominate in the beginning decades will pale, nevertheless, beside the impact that biotechnology will have in food production (and in medical fields) and beside the burgeoning knowledge base of nutriceuticals and their potential for health, for reducing the risk of certain diseases, and for well-being.

Genetic manipulation of food plant material and the cloning of animals will change the practice of farming. This impact on agriculture and the agricultural community and other technological consequences of biotechnology will be heavily felt by customers, consumers, and the food manufacturing industry alike.

The Achilles heel of these new technologies is this:

- They cannot be tested exhaustively on a small scale to determine if they are safe over extended periods of time (generations) in every conceivable use or under all conditions of operation.

Repercussions of these new technologies are being felt at this very moment and show no signs of abating. This is also an argument put forth by opponents of genetically modified foods, i.e., insufficient testing. New technologies will always bring unexpected results: socially, environmentally, and biologically.

Technological Issues Specific to the Food Industry

Some applied technologies have their greatest impact on the abundance and availability of food, its manufacture, its safety and wholesomeness, and its nutrition. Sparking controversy, for example, are the following:

- Raising genetically modified crops and husbandry of cloned animals
- Use of ingredients that have been derived from genetically modified sources
- Factory-farming or intensive agricultural practices and their consequent impact on the environment
- Inability to find an acceptable balance between sound conservation practices for agriculture or economic development and the loss of recreational forested areas and natural wildlife sanctuaries
- Acceptance of irradiation as a tool in the destruction of food pathogens in, for example, poultry products

Agriculture and Biotechnology

The biological sciences have always had a great impact on the agricultural industry. Biotechnology, the newest branch of the biological sciences, has brought new varieties of plants and animals with genetically modified characteristics to the farm as well as agricultural practices that were not possible before. Its impact could well exceed that which the computer and information technology have had on the food microcosm.

The new science has proven popular among certain sections of the peripatetic conference attendees judging by the vast number of conferences, congresses, symposia, and meetings that have been devoted to it. Scientists, for the most part, are enthusiastic about the promise of the field and are quick to promote this promise with the dreams it conjures up in their quest for new research opportunities and for support funds. Journalists have had a field day with the topic. It is one with political and scientific controversy and drama with the potential it promises for relief of human suffering. It has about it an element of science fiction.

Genetically Modified Foods

Food has been genetically modified for ages. Horses, dogs, cereals, and vegetables have all been bred for particular characteristics. The technique for doing this has been relatively simple and uncomplicated. One selected for those attributes one wanted in a plant or animal and mated or crossed it with a desirable trait in another plant or animal of the same species. The offspring, according to the Mendelian laws of inheritance, would possess the desired traits (it was hoped) which by further crossings would produce pure lines with the desired characteristics. It was a form of natural selection. A male and a female plant or animal were required, i.e., a sperm and an egg.

However, this was upset by biotechnology. Advances in this field introduced a new technique that was unconventional, indeed, was considered by many as unnatural. Up until this new innovation, only those species of plants and animals that could interbreed could be used to create new varieties of plants or of animals with improved characteristics.

The new technique is variously called recombinant DNA (deoxyribonucleic acid) technology, genetic engineering or manipulation, or just plain gene technology. DNA which represents a unique characteristic from

species (see Gasser and Fraley, 1992 for a general review of the procedure for plants).

Thus were born genetically modified organisms and from this the possibility, which has grown to a reality, of genetically modified foods. These are basic foods from plants with attributes derived from alien species of plants or microorganisms. The genetically modified organisms have attributes which they would never have possessed with traditional breeding programs. Thus consumers eat plants which have unique properties not obtainable through conventional breeding programs or they eat food products with ingredients that may have been obtained from plants so bred.

Conventional genetic modification is slow and cumbersome and without this new technology attributes from other species could not be introduced into other species. The classic example is the production of pest-resistant crops by the introduction of DNA from *Bacillus thuringiensis* which contains a natural pesticide in its genetic make-up.

Reactions to Genetically Modified Foods

The emergence of genetically modified food has elicited a wide range of reactions from the public.

- "Genetically altered food hard to digest, critics say" reads one front page headline (Abley, 1999).
- "'Frankenfoods' uproar grows" states another headline accompanied by a picture of activists dumping genetically modified soya beans outside the British Prime Ministerial residence. This story discusses the growing reaction against genetically modified foods that is prevailing in Europe (Greenwood, 1999).
- The use of genetically modified crops will lead to a 'farmageddon' (Lowey, 1999), whatever that means.

Journalists are having a field day coining new words as fast as the biotechnology firms are developing new hybrids of animals and plants. The list of screaming headlines could go on but those above are typical of North American reaction but modest in comparison to the large scale opposition in Europe toward genetically altered foods.

Some farmers have welcomed the advent of new crop varieties such

and politicians alike, fear the dominance of a few seed suppliers who are, in turn, controlled by biotechnology firms.

Public perception and understanding of recombinant DNA technology are poor. People are ambivalent about the potential value of biotechnology. They support, for example, the medical applications that this new technique shows promise of providing. Indeed, they are excited by the promise (and it is only that) of the elimination or control of genetic diseases, or by the hope of cell-culture of organs for the replacement of diseased organs, and by the hope of the genetic control of the human life span to permit a longer productive life. But these are personal concerns and people can make personal choices; they can choose to be treated by the new technology to prevent a life-threatening disease or not.

People do have serious concerns about gene manipulation, its safety, the ethics involved in its application in cloning of animals and, in particular, about the cloning of humans (not a food problem!). More seriously still, they have concerns about its safety in foods where they cannot make a personal choice. It is not a question of life or death. It is solely an issue of economic advantage bestowed to biotechnology firms supplying the seeds. The public is becoming very suspicious of safety as determined through science and by scientists. Many no longer have confidence in their veracity and impartiality (see Chapter 7).

The anti-genetically modified foods groups have been very active in fighting the new technology. Bioterrorists, following hard upon the heels of farmers planting the genetically modified seeds, have pulled up the seedlings. Anti-biotechnology feelings have been oddly diverse in their geographic distribution. The strongest distrust of, and animosity toward, genetically modified plants have occurred in Europe.

Anti-genetically modified food activity has been much less fervent in North America but is growing in intensity as many food processors have indicated their intention not to use genetically modified crops. This, they claim, they are doing in response to the demands of their customers/consumers. Throughout the rest of the world, opposition has been spotty but nevertheless is present.

Confusion for Legislators

Governments are confused, perplexed, and frightened by the technology and by the hubbub over the technology. There is no universal agreement

to derive a germ line of a natural product traditionally used in another country and patent its use worldwide? These are all questions which need to be resolved both within countries and between countries so that there is harmonization of laws.

At present, many governments are wavering on whether to adopt policies that are for or against the use of genetically modified plants and animals or on whether to require labeling of these foods. In some countries, the government's own scientists are split on what official positions should be. Dissenters against government policy find a ready outlet for their views among the many journalists eager to have a scoop. The lines are not clearly drawn.

The scientific community, as might be expected, generally supports the *science* of recombinant DNA technology. There is not, however, unanimity in the scientific community about either the long-term nutritional or environmental safety of the technology. Scientists are split. There are reputable scientists who support the use of genetically modified foods and those, equally distinguished, who are strongly opposed to their use. Main issues of contention center on long-term testing for human toxicity and long-term environmental safety.

Both sides, those for and those against the use of genetically modified foods and food products, are rarely short of qualified experts ready to speak for their side. Typical exchanges are illustrated in the following:

> *"To simply review such complex science [the regulatory review process for genetic engineering of crops], to accept the proprietors' data without independent validation – by the government or a university – is truly irresponsible."*

So commented E. Ann Clark, a crop scientist at the University of Guelph (quoted by Colapinto, 1998). The rebuttal by a Monsanto Canada representative, R. Ingrata, Director of Government Regulatory Affairs, as quoted by Colapinto was

> *"Why would we risk putting something on the market that we didn't believe was safe?"*

New Processes for Food Preservation

There have been some new developments in food processing but none

involved with these older technologies. For example, Appert's historic work introducing the practice of thermal processing has been furthered with many advances in other fields:

> Advances in packaging materials and technology have permitted lighter-weight materials, thinner-walled containers, and laminated films with unique gas transmission properties with which to package foods. The result has permitted such new developments as, for one, thin profile containers for rapid heat transfer and derivatives for modified atmosphere packaging for fruits and vegetables and minimally processed foods.
>
> Retorts have been redesigned. Now the contents of containers can be agitated during thermal processing and subsequent cooling to provide stable products of high quality with respect to flavor, color, texture, and nutrition.
>
> Better understanding of the theory underlying the kinetics of the heat processing of food as well as of the kinetics of the death of microorganisms has been developed. This has led the way to a logical approach to developing safer thermally processed foods.

Some of the innovative non-thermal technologies that are being developed to make more natural, nutritious, and safe products with extended shelf life are described by Mertens and Knorr (1992):

- High-pressure processing often in combination with mild heat treatment
- Biological control systems using antimicrobial enzymes and certain phytochemicals such as the phytoalexins
- High electric field pulses
- Oscillating magnetic field pulses
- Irradiation
- Intense light pulses
- Carbon dioxide treatment (packaging)

Other techniques that have gained success are ohmic heating, controlled/modified atmosphere packaging, and hurdle technology (Leistner and Rödel, 1976). Han (2000) discusses mathematical modeling to improve the functionality of packaging films that have antimicrobials incorporated in them.

Food Production and Processing Conflicting with the Environment

The production and consumption of food create waste, and waste pollutes unless it is treated. The production and distribution of food utilize energy and pollute. Debris from non-edible food (bones, for example) and from wasted food (plate remains) and packaging all create garbage which must be collected and recycled or used as landfill. Recycling utilizes further energy which also pollutes.

Historically, this has not been a problem. However, as the demand for food grew to feed the growing populations of the world and especially to feed the huge numbers of non-food-producing peoples, waste and pollution increased. Environmental and ecological concerns mounted worldwide as technology overcame the seasonality of food production, as people demanded fresh fruits and vegetables as well as tropical produce in all seasons and all to be available in volume. These demands did put strains on the environment.

Finding adequate solutions to the combined problems of waste and pollution are becoming more critical. The public, i.e., customers/consumers, is becoming alarmed:

- At the threat of global warming and the unknown climatic changes this might bring.
- At the loss of wilderness habitats and recreational lands due to the need for more agricultural lands. The loss of forested lands may also be a factor in global warming.
- At the pollution of rivers and streams from run-off from agricultural fields and by industry discharging untreated wastes into rivers.
- At the enormous, and steadily climbing, costs that cleanup represents to the economy.

All of this is caused by pollution. To be fair, not all of it is due to pollution in the food chain but much of it is.

Traditionally, waste has been treated by either of two methods:

1. Recycling. Collecting, sorting, and reusing the waste material (for example, the recycling of paper, glass, and aluminium cans) or converting some kinds of food waste into some usable byproduct (for example, animal feed, gelatin manufacture, fertilizer).

Both methods have advantages and disadvantages. Neither avenue is itself entirely free of secondary polluting.

Both are costly. Money, obtained through public taxation and/or producer fees (fees levied based on the weight of waste produced), is needed to set up the infrastructure to clean up the waste and pollution in some manner. On the profit side, outbreaks of waste-related diseases have been eliminated or greatly reduced in severity and number where the infrastructure to treat organic waste is in place.

Recycling has its challengers, especially to its value as an aid in improving the environment. An article in *New Scientist* has the startling title "Burn Me" (Pearce, 1997). In this, work is cited that clearly demonstrated that incinerating paper and some other wastes and utilizing the heat energy for the production of electricity is a far better, more environmentally friendly alternative than is recycling. Incineration, when done properly to avoid air pollution, reduced waste by as much as 75% and provided energy for processing. The ash which contains heavy metals from the print must be disposed of. Recycling involves de-inking, utilizes energy, and still does not prevent deforestation for more paper making.

However, the thinking of environmentalists has become so entrenched and governments have accepted this mindset that the more environmentally damaging procedure of recycling paper and other waste has in many countries become mandatory.

Clean Water, Clean Solids, and Clean Air

In most developed countries, traditional mixed farming has been challenged by the industrially managed intensive farm unit, more commonly referred to as factory-style concentrated animal operations. Poultry and hog production are often managed this way. Such farms can contain several hundreds to over a thousand animals each and can produce enormous amounts of waste.

These wastes (urine and feces) from the penned animals are washed out and held in nearby man-made lagoons. Lagoon waste is often then sprayed over neighboring fields. Two contamination problems can arise:

■ Groundwater run-off from fields sprayed with lagoon waste will ultimately drain into creeks and rivers. Here it has been known to cause algal blooms which can kill aquatic life. On a more cata-

■ Spraying disseminates the waste as an aerosol and ammonia in the waste is dispersed aerially causing this nutrient to be spread well beyond the borders of the fields upon which it is sprayed.

In similar fashion, pesticides and herbicides sprayed onto crops will eventually find their way into the soil, hence into the groundwater and finally into the water supply.

In order that these intensive agricultural practices that will be necessary to feed people in the future can continue to be practiced, these techniques:

■ Must be safe for, and acceptable to, the communities that they are located in or near; and
■ Acceptable to environmentalists concerned with the health and safety of the local ecology.

As waste treatment systems are expensive, much research is needed to develop methods for waste water treatment facilities that

■ Are economical for small rural municipalities to build and operate.
■ Are safe for the environment.
■ Reduce the amount of waste solids and/or create a profitable byproduct.
■ Produce clean water.

A small-scale pilot operation using human waste sewage employing what is described as a nutrient film technique was demonstrated by Jewell (1994). It produces clean water, greatly reduces solids, and has the potential to provide revenue-producing byproducts.

Welcome to the War Room

There is another technology that is important for surviving in the third millennium. Knowing what one's competitor is doing in the marketplace is important for the survival of any company. Getting such knowledge involves being in the marketplace to see

■ What products the competition has.
■ What new products they are launching

- How they are pricing their products; and
- What deals they are making with the outlets selling their products.

In short, one should know everything about one's competition. As the old adage says: "Know your enemy, the competition." The latest technologies allow manufacturers to accomplish this.

Traditionally, such snooping has been done fairly casually and overtly. Sales personnel visited stores, noted competitor's products, paced off facings, reported on prices, brought back promotional material, and talked with store managers. The information so obtained was returned to management. More covert activities involved the technical departments of the food companies:

- Competitive products were purchased, sent to the laboratory where they were cut, sampled, tasted, and analyzed. A crude formulation could be established from the analysis and a rough estimate of ingredient cost obtained.
- Technical and patent literature was reviewed to determine what research the competition had supported, to determine in what institutes it had been performed, and to find out what patents had been granted or were pending as a result of the competition's activities.
- Equipment sales personnel were casually questioned at trade shows. Who had purchased their equipment? What were these purchasers doing with it?
- Seating was arranged at a conference dinner beside the Director of Research for one's competitor or beside one of their senior technologists for overhearing any clue that could provide hints of the competitor's technical activities.
- At conferences, technical papers delivered by either the competitor's personnel or by research groups and graduate students supported by the competitor often revealed the competition's directions in their research activities.

Access to information regulations allows food companies to explore any forms, such as grant applications, that their competitors may have filed with their governments. Unrelated bits of information garnered from a multitude of sources, such as the few mentioned above, can be collated.

Benchmarking

Out of such low key activities, the management technique of benchmarking can be said to have had its origins. In a nutshell, benchmarking is closely examining one's own operations in all its aspects and comparing them with the leaders, that is, with successful competitors, in the same field. After careful comparison and analysis of the successful operations of its competitors, the company then applies the successful techniques to its own operations.

Benchmarking for a food company is, then, a process of constant improvement of services, of products, and of all areas of operations by comparison with the leaders in its industry or product category. Products can only be improved by comparisons with similar products manufactured by the competition. Points of comparison can be

- Quality characteristics,
- Composition and amounts of ingredients, and
- Price.

But also included is knowledge of the competition, who they are, and what their strengths and weaknesses are. It means researching the company's own customers as well as those of the competition. Once this knowledge has been digested and strategies developed on the basis of it, companies can take advantage to move quickly in the marketplace to gain superiority.

Such is the gentle world as it once was.

It's a War Out There: Industrial Espionage or Intelligence Gathering?

It is now called intelligence gathering. Companies practicing it professionally, for a price, can be found in directories listed under corporate intelligence or competitive intelligence. By another name or in times of war it would be called espionage or plain old-fashioned spying but it is legal, ethical, and done openly. From soft ball, the game of corporate intelligence gathering has become hard ball and very professional indeed. It has been estimated that globally collecting and analyzing information will be a business exceeding U.S. \$110 billion by 2002 (Lewandowski, 1999). The

is the same. It is the systematic gathering of as much information as possible about the competition from as many sources as possible and — this is the important issue — the intelligent use of this information to plan strategy against the competition or to counter the competition's strategy.

The tactics used include all those detailed at the start of this section (Welcome to the War Room) but they are not the only ways that information can be gleaned legally. Table 6.3 provides another look at some techniques that can be used. Some warrant a closer look with some further explanation as to describe their usefulness.

Much of what is displayed in Table 6.3 can be classified under the broad activity of networking, i.e., building up a broad association of contacts and using them as conduits for information. It is an innocent enough activity but the naïve individual may not be aware that he or she can be a source of valuable information that competitors can utilize.

Through networking at Chambers of Commerce, Alumni Associations, and philanthropic, fraternal, and professional associations useful competitive information can be found. Such groups have social activities and meetings at which prominent industrial speakers may reveal in their talks some of their company's business activities or afterward during informal sessions or receptions they may discuss or let fall some unguarded words about their company's programs.

Many companies encourage their staff to be active in these organizations and bring back any bits of information they hear or overhear. Professional intelligence gatherers certainly use these vehicles to ferret out snippets of information.

A company's own literature can reveal much about their present and perhaps future activities. Valuable information can be found in their annual reports, on their corporate Web sites, and in press releases. Besides the financial results that annual reports contain, the presence or absence of the names of company executives suggests that definite changes have been made. Who was hired (new executive names in the report) or who was removed or transferred (voluntarily or not) might indicate strategic policy changes that are being planned at that company.

Business newspapers record the movement of prominent business executives as they play their games of musical chairs, leaving one company to join another. The reputation and activities of an executive in a previous company can give some direction as to how that executive will operate in his/her new role. In a similar vein, help-wanted ads requiring individuals

Table 6.3 Sources of Information by which the Marketplace Activities and Strategies of Competitors can be Analyzed to Prepare Counterdefensive Movements

Source of Information	Specific Activity or Information
Print media	• Executive moves noted in business newspapers and trade journals • Press releases issued by companies • Help-wanted advertisements • Scientific and technical journals in which research papers indicate sources of funding and research topics
Corporate publications	• Annual and quarterly reports which often record personnel changes and policy directions • Corporate Web sites
Conferences and trade shows	• Exhibition booths at trade shows which demonstrate latest equipment and products • Suppliers often (in)advertently reveal activities of other companies as a sales ploy • Speeches made by senior management or research papers delivered by technical staff or by research groups supported by companies
Pro Bono activities	• Dinner speeches by senior executives at charity events, Chambers of Commerce, fraternal organizations
Government access to information	• Information and forms filed with various government agencies • Grant applications requesting cooperative research ventures • Patent notices
Company-sponsored social events	• Cocktail or other sponsored receptions or dinners • Sports outings such as golf tournaments

congresses and trade fairs or for some other celebratory occasion. Again, only to the very naïve are these entirely innocent affairs They are carefully orchestrated affairs. The hosts arranging them:

- Very selectively organize foursomes for golf.
- Plan seating arrangements at dinner tables so that maximum benefits for information gathering are had.
- At receptions important guests are targeted. That guest will always have a conversationalist present to keep the guest occupied. It is not by chance that at conference breakfasts, lunches, and dinners, representatives of a company rarely sit together but scatter widely among the other tables. Such mealtime conversation can be very revealing.

For the receptive company there are ample opportunities to gather information about what the competition is doing. It has even become very "high tech" with the reported use by some companies of satellite imagery to obtain information about a company's activities.

The food business has been, and will remain, very competitive. It is a business with comparatively low profit margins. Volume of sales is, therefore, very important. The competition is keen in the marketplace, but also in the boardrooms and senior management ranks. This competitiveness percolates all the way down through the various administrative and manufacturing levels.

Throughout the third millennium, it will not be enough to merely cut and examine one's competitors' products against one's self-manufactured products. Plotting successful strategy in local and national marketplaces as well as in various international marketplaces will require competitive intelligence gathering for both large and small companies. The tactics of carrying out the strategy have a greater chance of success if the intelligence that has been gathered is as complete as is possible, accurate, and diligently analyzed and interpreted.

Summary

Technology has increased the pace at which business is conducted. This has increased the speed with which business decisions must be made on the available data. Often the data on hand are incomplete. This puts heavy

of the beast that was to make business life easier has brought its own problems. The comment by Heidigger, an eminent philosopher of science, at the start of this chapter is certainly apt.

Understanding the marketplace and the players in it has become an important facet of doing any business. In the very competitive food business this knowledge and understanding of customers, consumers, competitors, retailers, and the variety of marketplaces are invaluable. Very specialized tactics must be used to analyze every move that companies make in the marketplace, in their organization, in their business and development activities, and in their government-related activities. Each datum must be carefully interpreted with all the other bits of data to understand the marketing strategies of competitors. Even the activities of employees at conferences, receptions, and social and fraternal associations must be scrutinized for whatever information can be extracted for usefulness to its competitors.

At the other end of the food microcosm an accord must be reached among:

- Those vociferous and influential opponents of the genetic modification of organisms
- Government regulators beset on all sides by issues of long-term environmental and toxicological safety concerns
- Chemical companies with control of the seed companies and who are the owners of the patents and
- Scientists who see the study of gene manipulation and its social, ethical, or environmental implications as research opportunities for grant applications for money.

It is an accord not soon to be won.

Chapter 7

Food Safety, Risk, and Quality

Salus populi suprema lex (The people's safety is the highest law).

Anonymous

Take calculated risks. That is quite different from being rash.

George S. Patton

To be alive at all involves some risk

Harold MacMillen

"All I Ask of Food is that It Doesn't Harm Me"

Michael Palin

The chapter title, the quotations, and section heading describe food safety, risk, and quality concerns that must be faced in the new millennium, i.e., the lessening of the risks related to the consumption of an increasingly diverse food supply by an increasingly heterogeneous and sceptical pop-

The risks associated with eating cannot be eliminated completely (cf. MacMillen above). Food risk can only be reduced to a reasonable certainty that the consumption of a particular food will not be harmful to the great majority of individuals eating it; it may be harmful to a very few. That there is this remaining small uncertainty is difficult to convey to customers and consumers alike.

Food stores can now provide a wide variety of fruits, vegetables, and meats from around the world as well as new food products developed by local manufacturers. Unfortunately, new foodborne hazards are also becoming apparent in the marketplace.

New variants of known food pathogens, for example, *Salmonella enteritidis* phage type 4 and *Salmonella typhimurium* DT104, are being discovered as causes of food-related illnesses. They are more virulent variants or have developed and increased virulence or have an increased resistance to common antibiotics. Certain microorganisms, such as *Listeria* and *Campylobacter* spp., have developed pathogenic properties. Prions, a newly recognised category, are hazards to human health. They are believed to be naturally occurring entities in the animal body whose function is not clearly understood. As the result of certain animal husbandry practices, they can be transmitted by meat products to consumers and to handlers of meat products to result in certain neurological diseases.

An aging population has created a widening age spectrum with increasing numbers of the very elderly. Both young children and the elderly are known to be more susceptible to food infections. As an added complication, the growing population of senior citizens results in greater numbers of people with age-related diseases and medical technology has created a growing number of people who are immunologically compromised.

With a world population expected to double every 40 to 50 years, one can anticipate that

- The number of mouths to feed with nutritious wholesome and safe food will increase.
- The number of people with specific health-related reactions or sensitivities to foods will increase.
- The number of people with special nutritional needs will increase.
- Diverse changes will be seen in the causative factors in food hazards of public health significance.

safeness among the food consumed, one's diet, one's lifestyle, and one's health been explored. Certainly the lack of certain micro- and macronutrients in the diet has been recognised. But that some components of foods could be factors causing or contributory to many diseases and that other foods contained factors preventing many diseases was startling (Cowley, 1998; Caragay, 1992; Cohen, 1987). The association of food with diseases such as cancer, coronary heart disease, and the variant Creutzfeldt-Jakob disease is recent. All of these are diseases with long dormant periods before they manifest themselves in their victims.

That food (and one's diet) could be dangerous to one's health raises issues for food safety that are yet to be resolved and introduces new concerns about the issues comprising the concept of safety that need to be faced. Suffice it, however, to introduce at this stage, in general terms, what these issues are.

- Risk and risk/hazard assessment are concepts that are poorly understood by customers and consumers. Nor are these concepts clearly understood by policy makers.
- There never have been tests performed with which to determine the safety of mankind's traditional or conventional foods. Indeed, there are no such tests. Satisfactory long-term nutritional, toxicological, and environmental/ecological studies have never been conducted for these foods. Their safety has always been assumed.
- Therefore, there are no tests with which to test or compare, with traditional foods, the long-term safety of, or the long-term risks associated with new and/or novel foods that come from exotic places, let alone new, genetically modified foods. Foods new to most North American and European palates that are available from around the world and particularly the new genetically modified foods have not been rigorously assessed for *long-term* ecological, environmental, or toxicological safety.
- Satisfactory methodologies to determine the safety of foods for *all* peoples against *all* known hazards are lacking.
- The trust of customers and consumers alike in the opinions, judgement, and wisdom of food, health, and safety experts has been badly shaken by their many reversals of opinion.

Nevertheless, unbiased, responsible intelligence respecting the risks

Food safety and risk cannot be decided only by scientific criteria as food, nutritional, and toxicological scientists would like to do. It must also be understood on emotional, political, social, legal, and economical levels. Without this broader understanding there will never be a consensus of opinion on the safety of food or the risks associated with food consumption.

Risk: Its Assessment and Communication

Customers and consumers do not have a rational outlook on risks associated with their food supply. This is the scientist's criticism and lament. Indeed, customer/consumers can be emotional, even irrational, in their concerns about food safety (i.e., unscientific in science parlance). On non-food topics where social, cultural, and traditional customs do not dominate or intervene, customers/consumers are less concerned about risk, and can even be defiant about taking risk. On food subjects, concerns are high.

Policy makers have one answer to risk: Eliminate it. In short, they legislate against it wherever they can. They put a limit on the amount of the hazard permitted in a food or they ban its presence completely from foods. This is a solution that is satisfactory to no one, for instance:

- It presupposes that there are methodologies to detect the presence of a risk factor and that all risk factors are recognisable as such. First, one can only do this with known risks. It provides no safeguard against unforeseen or unforeseeable risks. If the following incident or a similar one were transferable to a food incident, what HACCP (hazard analysis critical control point) or GMP (good manufacturing practices) programs, legislated or not, could have prevented the incident? There was a breakdown in telecommunication that blacked out a large part of southern Ontario. It was caused by a fire resulting from a spark generated when a repairman dropped a wrench.
- Legislation can only work against known or foreseeable hazards. It requires a list of all known hazards with limits to their allowed presence in foods. It also requires standard procedures for analysing for the hazards.
- This legislative solution requires that there be an infrastructure in place within government to examine all imports and at food

material or chemical added to a food must be proven to be both safe and to serve some benefit when added to a food. The onus of providing the safety data is on the requester. The danger here is that the evidence establishing safety is supplied by the petitioner. There is no assurance that the experimental trials were not designed to give greater credence to positive findings or that submitted data have not been selected to provide greater emphasis to positive findings and play down any negative results.

Risk

"No risk is the highest risk of all"

Title of an article (Wildavsky, 1979)

Scientists are the chief determiners of safety. They measure risk as a gambler would. For them risk is the odds of some toxic event happening in their experiments with a frequency greater than would be expected by chance alone. Scientists then lay claim to some significant event, not due to chance, as having occurred. They see risk at a statistical level.

The public does not understand risk at the statistical level. Why would they continue to smoke if they did? Or use cell phones while driving? Long-term studies on diverse populations of smokers have demonstrated that they have statistically higher odds for the development of lung cancer. Other studies have shown that odds for a road accident occurring are significantly higher when using a cell phone while driving. Individuals do accept risk voluntarily — but do not necessarily understand why — at the popular level, i.e., at the individual, or personal level. At the statistical level, nameless, faceless people die. At the personal level, the victim is known. Understanding is colored by personal or familial experience, social (peer) pressures, and other cultural accoutrements. A classical example of this dichotomy of concern for risk is exemplified by a radio newscast (January, 2000) about ice fishermen:

Four qualified persons, a Canadian Red Cross Society spokesperson, an aquatics safety expert, a government safety expert in the Parks and Recreation department and a safety engineer all advised ice fishermen not to drive their vehicles onto the ice during a January thaw. The ice was not safe to hold a vehicle until it was 40 cm thick, a figure all

184 ■ Food, Consumers, and the Food Industry

hole was eight to nine inches thick (roughly 20 to 23 cm thick). He knew from experience this was enough to support his vehicle. No deaths or accidents were reported.

Two types of risk, objective or theoretical risk and subjective or perceived risk, are recognised (Scherer, 1991). Theoretical risk for an event is derived through a rigorous statistical analysis of demographic studies, experimental models, and risk assessment processes. Subjective risk is the public's (or non-expert's) interpretation of this objective risk conditioned by personal value considerations based on social, cultural, and experiential elements (Scherer, 1991).

Safety: A Corollary of Risk

Safe is safe. There is no doubt, no little bit of danger.

Safety is a concept, a state of mind. It is personal. It is all-embracing. It wraps the mind in a warm, fuzzy feeling like a blanket. Risk is cold, calculated, statistical, and impersonal.

> *"Will central heating, lack of exercise, television and hyperclean food result in the decline of mankind."*

> Anon., 1999

The above statement does catch one's attention. Besides hyperbole there is a certain amount of truth in it also. Questions concerning the long-term implications of hyperclean or hypersafe foods are being examined by many scientists. Jay (1997) reviewed work done both in his own laboratory and by others which demonstrated the ability of harmless microorganisms in fresh foods to antagonise foodborne pathogens. He posited whether directing efforts to cleaner fresh foods was not endangering the public's health.

In short, is hyperclean or hypersafe food good for consumers, i.e., safe? Other researchers have suggested that hyperclean or hypersafe foods do not stimulate the immune system sufficiently; a system, they believe, that needs to be 'challenged' to function properly. Isolauri (reported in Shortt, 1999) commented that probiotic administration to young infants alleviated allergic reactions in them and checked the development of allergic symptoms. Therefore, hyperclean food may not be safe.

The public wants safe food. The problem arises when what "safe"

consumers require a basic education in nutrition and in food preparation techniques to select and prepare foods wisely. They need advice about the nutritive value of manufacturer's products, e.g., nutritional labelling, to buy nutritious products. It is inherently unsafe to leave nutrition solely to food manufacturers.

■ Toxicological and immunological safety. Food must be either free of hazardous substances, or these substances must be reduced to levels which are harmless in normal consumption, or the levels of hazards in the food will be eliminated in food preparation and handling.

■ Microbiological safety. The microbiological quality of basic foods and food products must be maintained at a non-hazardous and low level through distribution, selling, and home storage, or be reduced or eliminated during food preparation (but note above on hyperclean foods); and

■ Cultural safety. Foods which conform to all the above, must respect social, cultural, and religious traditions where they prevail. Foods must be available and so identified. This is an aspect of safety that is frequently overlooked by safety experts as there is no life-threatening aspect to it.

Risk Assessment

Winter and Francis (1997) summarised four elements of chemical food risk assessment that are applicable to new foods, including genetically modified foods:

1. Hazard identification. This process gathers and sifts animal and human toxicity data to examine the health problems that occurred (toxicity, birth defects, neurological effects, arrested development) and under what conditions they were produced.

2. Dose/response assessment. Here the likelihood of ill effects of human exposure to a chemical can be predicted after an effect of public health significance has been recognised.

3. Exposure assessment. An analysis of the preceding processes permits judgements to be made about the dose of chemical required to cause an effect in humans.

4. Risk characterisation. This process describes the magnitude, nature,

To set levels for toxic materials that may be found in foods, governments must determine (actually must be advised by experts):

■ What is a safe level for the expected pattern of consumption.
■ What is reasonable given the industry's ability to detect the toxic material, to monitor its removal through processing, or to reduce its level in food to negligible amounts? Safety must go hand in hand with the ability to analyse for the risk element.
■ A safe level for the consumer who is most susceptible to that risk.

There are interesting fallacies in the statistics of risk assessment figures (Matthews, 1998) that illustrate some of the difficulties understanding risk:

The risk of dying in a road accident in the U.K. is 1 in 8000 in any year. But these statistics do not apply to all classes of people. A young man is more likely — some 100 times more likely — to die or have injuries in a road accident than is a middle-aged woman. Someone who drives at 3 a.m. on a Sunday morning is many times as likely to die than is someone who is driving seven hours later. People with personality disorders are 10 times more likely to die in a road accident than those not so afflicted.

The data on which many risk assessments are calculated can be broken down into discrete compartments or subsets of like categories as the above example illustrates. It is then that the risk data can play tricks on the unwary. It may be found that one (or even some) subset(s) of the data demonstrate radically different levels of risk.

Table 7.1 presents a catalogue of factors that can influence the assessment of toxicity of a food or of ingredients that can be part of a food product. When comparable assessments are made respecting food risks, several such subsets of different toxicity levels can emerge. For example, in Table 7.1 if people eating the foods with the suspected hazard were divided into several subsets by particular characteristics, a statistician would no doubt find different critical levels of toxicity in each of the different subsets. That is, people of the same age, sex, height, weight, health status, and physiological condition might show a different risk tolerance than those with different characteristics. When there are lengthy response times before the appearance of any ill effects, any estimate of a risk in the population becomes incredibly difficult.

Table 7.1 General Overview of Factors Influencing the Assessment of Materials Toxicity in or Added to Foods for Humans

Source of the Factor	Factor Influencing Toxicity
Individual	• Gender of the individual eating the food • Age of the individual eating the food • Body size as measured by height and weight • Physiological status of individual (pregnant, lactating, convalescent) • State of individual's immune system
Diet and dietary habits	• Frequency of eating suspected toxicant • Level of suspected toxicant in food • Factors in the diet which protect against the toxicant • Presence of benign microorganisms in diet which can detoxify hazardous substances (probiotics) • Manner of food preparation
Nature of toxic response	• Short-term, acute response with symptoms rapidly apparent • Long-term response with symptoms not apparent until old age or in next generations

Decision makers look to science and scientists for help wrestling with the extremely ambiguous process of risk assessment. Scientists, however, cannot offer total certainty respecting the safety of anything and most certainly not of foods. Scientists can only demonstrate the nonobservance of harm (toxicity) within the limits of the experimental design of their experiments. They cannot extrapolate beyond these limits nor can they assess the impact on safety should the suspected toxicant have a long (years, indeed decades long) dormant period.

With the risk measured as best as can be possible with the tools available there remains the task of risk management (Chapter 8) followed by risk communication.

188 ■ Food, Consumers, and the Food Industry

> *Britain's Committee on the Safety of Medicines had announced that*
> *the latest types of oral contraceptives were more effective but were*
> *twice as likely as the previous formula to have side effects, that is, to*
> *cause blood clots. The result of this warning was that many women*
> *did not use the new contraceptive. It was estimated that 8,000 extra*
> *abortions were performed as a result of this reluctance. There was no*
> *estimate of the number of unplanned pregnancies nor of the greater*
> *hardships that resulted from either the abortions or the unplanned*
> *pregnancies. Which was the greater risk? The new contraceptive*
> *increased the risk of clotting from three in a million to 5 in a million.*

There, then, is the dilemma of publishing statistics to a public not familiar with assessing risk in this fashion.

Three basic assumptions continue to be made by scientists in communicating risk (Scherer, 1991):

1. Truth, especially objective TRUTH, is a product of science and science alone. This is the crutch of the scientist and the technocrat.
2. Scientists and their devotees, technocrats, are the only apostles of this Truth.
3. The public is a "malleable and a uniform mass of information receivers" (Scherer, 1991) who once informed of the TRUTH would, or should, accept it unwaveringly.

Scherer demolishes each assumption.

■ Scientists have their individual biases (see later). They make errors and they can be greatly influenced in the interpretation of their data.
■ Scientists firmly believe that only answers derived from scientific principles are rational but the public, on the other hand, may use other equally valid criteria to establish truths (for a further discussion, see Scherer, 1991). The public believes scientists are not the only appointed deliverers of any knowledge.
■ The product of science is indisputably an explanation of the material world. This is a far cry from the TRUTH. Science is far from being a viable tool to explain the human experience or the humanities.
■ The public is not a passive sponge eager to absorb the judgements

These false assumptions, credited to scientists by Scherer (1991), contribute greatly to what has been described as the public's "outrage factor" in risk assessment. Not having their value judgements concerning risk acknowledged or considered valuable by scientists is one cause for the growing rift between scientists and the public. Such outrage factors are ideal fodder for use by activist groups to raise as issues or by eager journalists investigating and developing the sensational science.

It is not in the scientist's nature to communicate absolute certainty on any topic and certainly not on the difficult subject of toxicity. Nor is it in the scientist's best interests to be certain. By admitting to, or even creating some reasonable doubt, and hence insecurity and fear, the scientist wins the funding to allay that fear through further research. Funding is what the scientist wants.

"Science knows only one commandment: contribute to science."

Bertolt Brecht, *The Life of Galileo*

Safe results do not always get reported in the popular press. They do not get the scare headlines that upset people enough to coerce their government representatives into providing further funding. Bad results, i.e., the possible health scares associated with some industrial practice, are all that are heard.

Extraordinary Food Risks

Not all food risks can be identified, assessed, and communicated. There are unexpected reactions that many people have or may develop to certain foods or components of foods. There is food terrorism, malicious vandalism, and extortion using food as a target. Control efforts are possible but unlikely to be widely adopted, or highly effective, or even desirable for customers if adopted.

Food Reactions

Many people are afflicted with allergies or have unpleasant, but not necessarily true allergic, reactions when they eat certain foods. Amongst my circle of friends there are those with reactions to nuts in general and

in particular). Such reactions to foods can range from life-threatening ones to simple discomfort or embarrassment.

The most obvious solution for these individuals is to assume the responsibility for the avoidance of the particular foods they are affected by. Control is entirely in the hands of the consumer. To have this control, however, consumers must be informed of the composition of the foods they eat. For this, they need proper and full disclosure of all ingredients on the labels of all processed food products.

The great danger for these individuals comes in places where food is not labelled or is incompletely labelled. For example, prepared foods are not labelled with their ingredients:

- In restaurants and other eating-out situations (company picnics, church socials, etc.).
- In butcher shops or delicatessens selling raw or cooked sausages and meat pies or other semi-prepared products made on the premises.
- In retail situations where wrapped goods such as candies (below a certain size) are sold loose in bulk.

In these situations of personal risk, safety can only rest with the consumer and the honesty of the vendor in describing the composition of the prepared food products being offered. When such information is unavailable or is suspect, consumers must exercise caution with their food selections.

Food Terrorism

There is a worldwide fear that militant terrorist groups can operate secretly to wreak havoc anywhere in the world to advertise or support their causes. Weapons have become more and more terrifyingly powerful and able to cause devastating harm. They are easily portable. The Oklahoma bombings, the U.S. embassy bombings in African, and the nerve gas attack on the Tokyo subway are prime examples of what terrorists can do.

Food retailers have experienced vandalism in their stores by radical fringe groups who want to publicise their causes, and by disgruntled employees. Retailers must devise ways to control such malicious acts within their stores whereby sensitive foods could be contaminated. Some solutions may be unpleasant and resisted by retailers; for example, control customers'

- Entrances and exits to food manufacturing plants are secured with guards to monitor passage of properly identified personnel or visitors only.
- All staff require proper identification with name tags and/or key cards to limit access to vulnerable areas of the manufacturing operation.
- On plant work floors, plant personnel are identified with color coded clothing so that supervisors can recognise when personnel are "out of place."
- Adoption of screening techniques to identify and assist troubled individuals and prevent their efforts at malicious damage.

Far greater danger from food terrorists can occur on a national scale. A nation's agricultural food supply represents a vulnerable target for attack with bio-weapons such as air dispersible plant and animal pathogens. Attacks upon the national food supply conducted by small groups of terrorists bent upon weakening a nation by crippling its food supply, destroying the agricultural economy, and endangering the health of its people are realities. Health scares, increases in food prices due to crop failures, or presence of plant pathogens can destabilise governments (cf. the dioxin scare in Belgium), cause riots among hungry peoples, or force governments to spend huge quantities of money to eradicate the disease and feed its people.

Terrorist activities are presently far beyond the ability of any nation to prevent.

A more frightening possibility is that these attacks need not be directed by extremists at political or ideological enemies (MacKenzie, 1999). Economic rivals could just as easily conduct such internecine acts of warfare against their competitors.

What Methodologies for What Hazards?

Man did not choose his food; man's food chose man. The source for this statement is lost. Its premise is sound. One can argue that traditional foods are safe only because by their use throughout humankind's history, all the people who were sensitive to these foods have been eliminated. Their sensitivities could not be passed on to future generations. Only those who were able to develop a resistance to available foods survived; they lived

that is, eat according to heritage. People with type O blood groups were hunters with acid stomachs more suitable for digesting meat, so the theory goes, while type A blood carriers were more suited to vegetarian diets.

It is a moot point how tradition and culture influenced people's sources of food and how the variety of traditional dishes based on these sources developed. How the early peoples determined which foods were safe and which poisonous is beyond comprehension given today's body of knowledge concerning the potential hazards in the plants and animals in nature. Traditional foods are deemed to be safe solely by virtue of an unwritten grandfather clause. These foods have never been tested by any rational system of methods used for toxicological studies today.

Many foods would not be acceptable today if tested by current toxicological methodologies. These are alcoholic beverages (biogenic amines in beers, Zee et al., 1981; a variety of constituents in wines, Tomera, 1999), vinegars, crops grown in soils with a high selenium content, and high salt foods such as some anchovies (12 to 15% salt). For a more complete listing, the reader should see Ames (1983) who reviewed dietary carcinogens and anticarcinogens and Diehl (1996) who documented natural toxicants, mycotoxins, and process-associated toxins in foods. Newspaper headlines (January, 2000) proclaimed oyster sauce and soy sauce contained a carcinogenic chemical "...shown to cause cancer in rats when fed at high doses over prolonged periods" according to a British health official. So much for the safety of traditional foods (Dawson, 1999).

Safety concerns have arisen for genetically modified foods and for the many new fruits and vegetables from countries around the world which are being sold in the developed countries. They have never been tested for long-term toxicological, nutritional, or ecological/environmental safety with acceptable methodologies to make valid comparisons with traditional foods. The question must be raised: How were the varieties of fruits and vegetables bred by conventional, i.e., sanctioned by tradition, techniques tested for either their short-term or long-term safety? Then, too, there are those varieties of fruits, vegetables, cereals, and oilseeds that were developed by techniques just as unconventional as those used for genetically modified foods. These foods were genetically modified by techniques employing mutagenic chemicals or nuclear irradiation.

Questions about the Methodology for Determining Safety

tolerated dose at which the animal does not lose more than 10% of body weight over its lifetime. These levels are far in excess of levels that human beings would have encountered in a normal diet. In the U.S., for example, doses to produce tumours in animal studies are several thousandfold higher than human exposure might be (Winter and Francis (1997).

A dose-response curve for animals developed from such animal study has no relationship to human beings. It contains no data in the low dose range of greatest importance in human feeding. In addition, such tests are performed on genetically identical strains of animals — humankind is not genetically identical — nor can animal or microorganism studies be completely reliable models for human toxicity (see Table 7.1). Humankind also does not eat a uniform diet as do the test animals used in such studies.

Scientists cannot (or ought not) extrapolate beyond the upper and lower limits of their experimental data. They can interpolate only between the extremes their data represent. This restriction applies to all experimental work including toxicity studies. Therefore scientists can never know or make conclusions about the effects of very low doses beyond their low points over very long periods of time, i.e., over lifetimes or even over generations.

Animal bioassays designed to observe toxic effects at high dose levels will not, of course, reveal potential effects (some of which may even be beneficial) at low dose levels. "…(L)ittle thought (has been) given to posing the reverse question, namely can food chemicals provide benefits and how such benefits might be quantified" (Lindsay, 1998). Increasing numbers of non-nutrient chemicals in foods have been shown to have beneficial effects in living cells, especially in low doses but are toxic at high doses, for example, selenium (see, for example, Burk, 1976; Reilly, 1998), arsenic, and alcohol as well as more toxic chemicals.

Lindsay has other criticisms of the present thinking regarding safety methodology:

- Chemicals with presumed toxic properties are considered risky without regard to other protective factors that may be found in foods which may prevent the toxicant's activity.
- There are thousands of primary and secondary metabolites in plant or animal foods whose toxicity has never been examined. If these metabolites were subjected to present testing procedures they might be branded as toxic.

Lindsay's argument is that the whole diet must be considered when evaluating toxicity.

Perhaps it is reasonable to rethink the methodologies used to assess food chemical risks.

From Professional to Amateur Food Handlers: Growth of Risk Factors

Food manufacturers have developed their own strategies and methodologies to safeguard their products from hazards accompanying any raw materials and from errors in their processing.

- They purchase from trusted and reputable suppliers and co-packers with whom they have established a history of quality materials safe from hazards.
- They investigate new suppliers and co-packers prior to dealing with them.
- They buy food materials against known or recognised industry standards or to standards they have developed which ensure the products they manufacture are safe from hazards of public health significance.
- They require signed affidavits from their suppliers that declare the wholesomeness and safety of the ingredients supplied and/or that absolve them of any losses or harm in the use of the materials supplied.
- Manufacturers also conduct on-site inspections of suppliers. They analyse their suppliers' products for obvious defects and for known or suspected hazards in food supplies that might be associated with the geographical source or character of the raw material.
- They contract with growers for produce that will be grown according to agronomic practices that they (the manufacturer) have approved. Staff agronomists are allowed to spot check contracted fields.
- Co-packers are regularly inspected during the processing of their products for adherence to contracted standards and to other processing, warehousing, and distribution codes that are designed to maintain the product-in-process to be safe, and free from hazards until it reaches the retail level.

- They maintain facilities in-house to monitor a product's "designed-for-safety-and-quality" specifications and to monitor safe process conditions according to principles of programs such as GMP (good manufacturing pactice) and HACCP (hazard analysis critical control point).

All or most of these methods are used by food manufacturers to ensure the safety and wholesomeness of their products from the field to the retailer or distributor. They are the manufacturers' only tools for protecting safety and quality.

There is no individual or collective absolute guarantee of safety. A certain amount of risk respecting the safety of food must be accepted. All the calculations, expert opinion, and preventive practices that support the calculated risk are designed to reduce risk to as low a level as possible. As Patton says in the introduction it must be a calculated risk. But it cannot ever be zero.

The products now go to the retailer.

The Disastrous Loss of Safety and Quality from Store to Home

The degree of safety designed and built into food products by their manufacturers begins to deteriorate with the retailer right on to the customer. It continues with any further preparation done to the foods by customers as they prepare meals within their households.

It is at the retail level that

- Part-time, unskilled, and often untrained labor stock shelves with time- and temperature-sensitive food products, i.e., fresh produce, refrigerated goods, and frozen products. This labor force is often without any realisation of the importance of the care they should take in what they are doing. Consequently, temperature-sensitive products are displayed above the safe temperature-holding capacity of the equipment during heavy customer traffic periods.
- Fresh produce, refrigerated goods, and frozen products often sit on store loading docks or in the aisles of stores while heavy customer traffic in the store is attended to.
- Untrained and inexperienced food handlers stock and attend the deli counters where prepared foods are displayed for immediate

products in them, often replacing them back on the shelves only to retrieve an unopened one. Customers also sample product at deli counters and open salad bars and return uneaten portions to the trays and bins.

■ Terrorists, mischief makers, or extremists from vested interest groups, for example, anti-factory farming groups or anti-irradiated foods groups, have ample opportunity to vandalise produce.

The retail level needs to assume greater responsibility for safeguarding not only the quality of the food but its safety from their central warehousing facilities to the retail counter. Their role is not only as purveyors of food; they have a responsibility for the safe and secure storage and display of food.

Most supermarket employees have had little or no training in the handling of fresh produce and refrigerated meats. They have never had any training in food service handling. Often they are young, and/or part-time, and/or minimally educated in food service handling and unfamiliar with or unaware of the hazards associated with cooked food handling.

In one supermarket (American Midwest), I witnessed a worker remove a tray of meat pies from an oven situated in the meat preparation area. The employee placed the cooked meat pies on one end of a cutting table where a butcher, at that moment, was trimming raw meat at the other end, roughly 2 or 3 meters away. The employee then left the area to retrieve a dolly already carrying wrapped uncooked meat products. She placed the uncovered cooked meat pies with these and pushed the dolly out to the fresh meat counter. The raw meat products were unloaded. Then the cooked meat pies were set into the deli counters.

This activity is totally unacceptable and should the trend to the deli counter within a supermarket continue food retailers are courting disaster. Retail outlets have to better educate their employees in the need for proper food handling techniques, demonstrate them to their employees, and maintain some system of quality control over these food service operations especially where there are self-serve systems. With the eating-out experience growing and with the customers' desire for fresher, i.e., less processed foods, retailers must handle much more fragile products such as refrigerated, modified atmosphere-packed produce; refrigerated, minimally processed products; and refrigerated processed foods of extended durability.

Food Safety in the Home

Worsfold and Griffith (1997) reviewed the food safety habits of 108 candidates drawn from various audiences. The subjects prepared dishes according to recipes using ingredients frequently cited in food poisoning outbreaks, for example, a cold chicken snack, a baked egg dish, and a minced beef sausage dish. They lost points for all the inappropriate and hazardous operational steps they took from the shopping center to the final preparation of the dish and won points for appropriate control measures. The authors noted incidences of no hand washing after handling raw chicken, improper cooling of cooked products, poor temperature control of refrigerators, and cross contamination between cooked and raw food storage.

Daniels (1998) in an informal survey of 106 households in the U.S. and Canada found that a number of critical violations (ones that could result potentially in illness or injury) and major violations (ones that could be probable contributing factors in food illnesses) occurred in the home. His conclusion was that food preparation practices in most (99%!) of households surveyed would not be acceptable according to a food-service evaluation system. A follow-up study was performed in the second quarter of 1999 with 121 people in 82 North American cities (not the same candidates as in 1998). In this study it was reported that almost 75% of candidates practised unsafe food handling procedures (Audits International's Home Food Safety Survey, 2nd quarter 1999, http://www.audits.com/Report.html) (Anon., 1999a).

Both these studies point to the growing need for responsibility within the home for control of food poisoning episodes. Customers must assume responsibility and liability for their own ignorance in the handling of food. Their abuse of food starts, as seen in the above, where the abuse often encountered at the grocery store left off.

■ The time/temperature-tolerance of fresh, refrigerated, or frozen foods is strained in the interval between the store and proper storage in the home. Customers often proceed with other shopping after storing their groceries in non-ideal conditions in their cars in mall parking lots. Lengthy travel distances from urban centers to suburban malls add to the abuse.

■ Most customers do not know the temperature in their refrigerators. These temperatures are often set too high for good refrigerated

- During food preparation, proper food hygiene is rarely maintained. Cutting surfaces and preparation surfaces are not cleaned and sanitised between handling of raw and cooked food. Hands are not washed between handling of raw and cooked products or after distractions such as telephone calls and interruptions caused by family.
- Oven temperatures are often inaccurately maintained. Finish temperatures of products may not be reached and held for an adequate time for safe heating.
- Prepared foods are often cooled improperly or left at room temperature for too long.

Customers and consumers alike are becoming increasingly ignorant of food, food preparation, and nutrition. This ignorance is clearly demonstrated by an earlier U.S. survey conducted by Williamson et al. (1992). These authors concluded that proper food handling is lacking in the home and consumer education is necessary. Consumers often lack skills in the hygiene of food preparation as well as in personal hygiene. The likelihood, therefore, that they can be entrusted to safely handle, store, and prepare the foods they or their families eat becomes more problematic. The serious question arises: Where does personal responsibility begin?

If such ignorance of food is becoming the norm for customers and consumers, the question becomes: Must manufacturers and retailers take more and more precautions to not only protect themselves from litigation but also to protect the public from their own ignorance? The answer is "No." This is an impossible task for manufacturers and retailers to carry out alone.

Customer/Consumer Responsibility

Customers and consumers must bear responsibility for food hazards which they are the authors of. Food hazards encountered because of careless storage and handling of food or by improper food preparation in the home can only be blamed on the public's ignorance about food and its proper and safe preparation.

There can be several reasons for this ignorance:

1. Teaching of home economics (for both sexes) has either disap-

secondary school level and largely only by faculties or departments dedicated to these subjects.

2. Lifestyles have changed drastically. Fewer meals are eaten at home with all family members present. More women have opted to remain in the workforce and pursue their own careers. People lead much more active lives, they eat at different times to accommodate personal activities, and so require frozen, prepared meals.

3. Less food preparation, therefore, is carried on in the home. Children and teenagers have less opportunity to assist and be instructed, however vicariously, in both food preparation and in meal planning. The skills and knowledge respecting proper handling of foods and the nutritional lore of parents and grandparents are slowly eroding.

4. Food manufacturers see great product opportunities to capitalise on these changes in the lifestyles of their customers. They have developed products that require no or minimal preparation. Only reheating is needed and instructions for this are placed on the packages.

5. Fast food outlets with their take-out capability and wide variety of ready-to-eat food selections negate a consumer's need to be knowledgeable about foods or nutrition. Food retail stores compete with these food establishments by creating their own take-out deli food counters. Parent are even spared the task of, and knowledge for, making lunch for their children as packaged prepared lunches have become available.

In short, there is no need for customers or consumers to know anything about food, its nutritional value, its proper storage or handling, or its preparation. One does not even need to read the cooking instructions on packages of ready-to-eat products if what the technologists tell us is coming. Microwave ovens will be able to scan the directions and prepare the food for the consumer according to the manufacturer's instructions.

Another trend that might influence both food and nutrition in the new millennium is broadly described as the home meal replacement trend (Hoch, 1999). It is certainly not a worldwide trend and can only be fully understood in North American food habits. Home meal replacements are food preparations that require minimal processing at home yet provide good nutrition and are the equivalent of a meal. They can be complete

obstacle to good nutritional practice; that and pressures of work and the desire to use time in more creative and hedonistic ways. In addition, ignorance of cooking is becoming a contributing factor to the popularity of home meal replacement products.

Hoch (1999) points out that this trend to home replacement meals will put an end to home baked food "just the way mother used to make it." A whole generation of children will grow up with no knowledge of how to cook because they were never taught by example at home or learned at school.

The above developments should not be viewed as a technological success. If anything they have potentially catastrophic consequences for the health and welfare of nations.

Future Implications

What the implications of this for the future are is hard to say. If home meal replacement becomes an integral part of an individual's meal planning — and it has for many people — the result could be any or all of the following:

■ Generations of children will grow up with little or no cooking skills. Only those with a hobby interest in cooking will venture into "cooking from scratch." Many cooking skills will be lost.

■ Younger generations will come to rely on home cooking as merely the assembly of prepared home meal replacement items and also to rely on fast food meals (Hoch, 1999). The ability to choose components will allow a strictly hedonistic approach to choices for foods, choices which are not necessarily healthy ones.

■ Knowledge of nutrition and concern for nutrition will move from the customer/consumer to the food manufacturer. Food manufacturers will have an increasingly dominant role in nutrition, selection, availability, and cost of foods eaten.

■ The family meal time as an opportunity for communication within the family may well be seriously eroded. Individuals in the family can eat when they want according to schedules they have set, and eat what they want according to their tastes.

This depiction of doom and gloom to come may appear to be an

and accept liability if they are at fault. Putting this idea across to these entities will be a monumental task in the new millennium.

An Erosion of a Belief in the God of Science and Its Apostles

"Everything I like is bad for me. But if I wait long enough, you food experts change your minds." This came from two friends of mine recently. A lament like this is not uncommon; others have voiced it with very similar words. For example, the following claims and rebuttals have appeared in newspapers and magazines over the past few years:

- Foods with high cholesterol have been bad and good by turns when exogenous cholesterol was found to be not as guilty as suspected in heart disease. Eggs, one common food, were bad but then became a good source of nutrition in moderation.
- Coffee and chocolate have had a checkered history. Coffee was a cause of pancreatic cancer but then it became a valued nutriceutical. Today, it is a cause of arthritis. Chocolate, a food prone to make its eaters tired, violent, suicidal, depressed, and suffer from migraine now has health-giving properties and may even be, it is suggested, an aphrodisiac.
- Butter, an animal source of fat, was condemned as a dietary fat; oleomargarine was good. Now consumers are told that the short-chain fatty acids in butter are good and that the trans-acids in oleomargarine are bad. Manufacturing practices for oleomargarine were changed to remove or lower the content of trans-acids in some oleomargarines. Consumers are now not sure what to believe about their oleomargarines.
- At one time unsaturated fatty acids were very desirable in the diet. There was a race to put more and more unsaturated fats into products. Then too much unsaturated fat in the diet was found to be bad for one's heart.
- Saturated fats, even those of plant origin, were once condemned as being as bad as animal fats in the diet. They now have been shown to be not harmful. Nevertheless, one can still find food labels proclaiming that the product contains no saturated vege-

but another study found that resveratrol in red wine offsets some of the ill effects of smoking.

■ Fiber was a protective factor in preventing colorectal cancers. Everyone was encouraged to increase the amount of fiber in their diets. Then another epidemiological study suggested it did not have any protective action (Fuchs et al., 1999) or if it did, this might be genetically linked.

■ Once oats with its fiber was lauded as a potent factor in lowering blood cholesterol. Then it did not. Lawsuits have been waged over this claim. But yet another report in the press of a recent study by researchers tells the public that a diet rich in oats not only lowers blood cholesterol but regulates blood sugar and lowers blood pressure.

■ β-carotene, and hence foods high in it, helped prevent cancer of the lung but then it did not.

The June 28, 2000 issue of *The Gazette, Montreal,* had a front page headline (Spears, 2000): "Ottawa Discovery: Virus destroys cancer tumours." The next day's issue of the same newspaper also had a front page headline (Derfel, 2000): "No cancer miracle, MD warns." Both articles were reporting on the same topic. Such flip-flops as this 'cancer cure' and those above of supposedly scientifically based opinion are confusing to the public. They shake the public's faith in scientists.

More serious incidences of the fallibility of scientists have challenged the public's belief in scientific (expert) opinion.

■ Biologists praised the virtues of introducing alien organisms as pest control agents — biocontrol it was called — until the introduced organisms began to become themselves pests as they invaded new biological niches and competed with local beneficial organisms.

■ Chlorofluorocarbons were hailed as chemicals of great benefit to humankind and harmless. Then it was discovered that they were very stable in the environment, that they were virtually indestructible by soil microorganisms, and that they damaged the ozone layer;

■ Scientists declared that lead in gasoline was not a hazard in the environment. It was not a contributing factor to lead accumulation in the body. They were wrong; it was a major contributor to lead

- DDT was touted as a valuable and safe chemical in the service of humankind. Then the damage to the environment, its persistence in the environment, its harm to birds and animals, its accumulation in human body fat, and its interference with liver function were reported. Nevertheless, some experts denied these results and resisted its banning. Dioxin was similarly claimed to be a safe and valuable chemical for human use.
- There is growing evidence to support the claim by reputable scientists that the outbreak of bovine spongiform encephalopathy in the U.K. was the result of a scientific experiment designed to breed better cattle that went tragically wrong. The disease, in its turn, spread to humans as variant Creutzfeldt-Jakob disease (Barnett and Wintour, 1999).

Only a strong public reaction, i.e., public outrage, concerning these expert misjudgements forced retractions before anything was done.

Faulty Delivery of Science to the Public

Currently scientists call a press conference when they have developed some new technology or made some remarkable discovery, even before they have submitted their research findings for peer review and publication. There are several less than altruistic reasons for this precipitous action:

- They want quick recognition.
- They want to pre-empt a rival group of researchers (who will no doubt contradict the findings of the former group as premature or irresponsible).
- They are looking for grant monies for further research.
- They are looking for venture capital in order to hive off a small company.
- They want government approval for whatever drug trials they would like to carry out.

Later when their often incautious remarks have been worked over by the media, more cautious statements emerge. These later statements are heavily censored by the public relations departments of their universities or by supporting companies, pointing out that they are only preliminary

perhaps even a cure (italics mine). They were seeking funding and government approval for the drug. They admitted that another 5 years of testing and research were required before any definitive results would be available.

Scientific Differences of Opinion

The public is amused at or more likely exasperated with the opinions and counter-opinions presented by experts. "Beset by the 'carcinogen-of-the-month club' people see their eternal verities turn into mere matters of opinion" (Wildavsky, 1979). Certainly there is no disputing the observation that qualified experts can each present different interpretations of the same experimental evidence. When experts do not agree, how are the differences to be resolved? To whom is the public or government to turn for guidance?

Technologists and scientists acknowledge that as the frontiers of knowledge expand opinions and interpretations will change as more data accumulate. Theories and opinions will be reinterpreted in the light of the new findings. Nevertheless, the general public, all of whom are customers and consumers of the products of science in some manner, are neither technologists nor scientists. To use the vernacular, they are tired of being jerked around, and confused or frightened by scientists who are spouting expert opinion without all the pertinent research performed or who are following personal agendas or vendettas.

The proper forum for such publication of new findings is in publication of peer-reviewed journals or scientific conferences, not in press conferences.

Possible Reasons for a Distrust of Scientists and Expert Opinion

I had several occasions to have dinner with James Delaney, he of the famed, or infamous depending on one's point of view, Delaney clause of U.S. food regulations. On one occasion, in the members' lounge at his club we talked over coffee and brandies about the Delaney clause and how it came about. He related the following story to me:

room. The next morning when he entered the hotel lobby, he found
one of the scientists pacing up and down, waving his arms about, and
talking to himself! He decided, at that moment (so he told me) that
he would draft a piece of legislation that was so ironclad that when it
was changed (and he expected it to be changed), the legislators on
the advice of experts would know what they were doing and why they
were doing it.

There is said to be a little madness in all people. Was a "mad scientist"
behind the Delaney clause?

The mad, or certainly very disturbed, scientist of cartoons, fiction, and
the sci-fi cinema is an all too well-known caricature to the general public.
When the answer to the clue "like a scientist" in a Sunday *New York Times*
crossword is "mad," scientists have certainly become objects of some
ridicule and disrepute. What else would journalists and subsequently the
activists call genetically modified foods other than "Frankenfoods" if they
had not had Mary Shelley's much misunderstood *Frankenstein* as a model?

Burke (1999) suggested four possible reasons for the public's distrust
of, or lack of confidence in, expert opinion respecting foods and diet.

1. "…scientists have sometimes been too influenced by commercial
   or political pressures…" Interference in science has been demonstrated historically by the Roman Catholic Church and by the
   Russian Communist scientific oligarchy that interfered with genetic
   research.
2. Consumers, that is, the general public, do not understand the
   concepts involved with risk and risk assessment, nor do they
   understand the degrees of risk which are often used by scientists.
   These concepts are poorly explained to these same consumers.
3. Pressure groups confound the issue of what is truth for consumers
   by enlisting the support of scientists who have their opposing
   views. They, then, put forward conflicting evidence to support
   their points of view respecting safety, toxicity, nutritional value, or
   social acceptability.
4. The context in which a risk is presented greatly influences its
   assessment. A newspaper headline "Healthy Food Makes People
   Ill" (Thornton, 1999) is startling and frightening. It is not until the
   second paragraph that the reader learns this is about people who
   have sensitivities to foods. Or the headline I read once that the

Risk and risk assessment have already been discussed with respect to the customer's and consumer's lack of understanding. The remaining issues raised by Burke will be discussed in greater detail now.

Fraudulent, Unethical, or Questionable?

Scientists are biased. The purity of science is, or can be, more than slightly sullied. These are observations that the public is making. This loss of respect for science requires exploration to understand its causes and to determine the best avenues through which science can regain the trust and confidence of the public.

In the Beginning: A Wedding of Convenience

Declining government funding in many countries resulted in universities becoming strapped for money. In the U.S. this crisis dates from about 1960 when the U.S. government supplied roughly 65% of university funding and private industry approximately 33%. Government funding fell sharply. In 1995, the government only provided 36% and private industry supplied 59% (Moore, 1998).

In Canada, during this period, a government policy was instituted whereby funding was to be closely tied to projects that were application-related and hence valuable for industry. In the U.K., a similar policy was adopted. Moore, then president of Sigma Xi, the Scientific Research Society, observed that the research community, in which he included the university community, would have to adjust to a new funding aim more responsive to the needs of industry. This supports Burke's (1999) thesis that science had to bend to political pressures.

In the belief that economic growth would be sparked by a more technologically driven society, governments encouraged their universities to solicit industry for research funding. In addition, by directing funding policies toward applied technologies, governments have shown a direct bias toward research with an immediate application for industry rather than for pure research. The research must benefit industry or the research proposal does not get recommended for funding.

Professors (with their universities' permission) responded by selling their services to private industry. In this manner, they got the necessary funds to maintain their research programs which supported students

and with cheap graduate student labour. On another side, there were the pharmaceutical, chemical, food, agricultural, medical, and analytical supply companies as the white knights who could possibly save the universities. And there was government with tax concessions for industrially sponsored research and development and grant monies tied to applied research.

Private industry had their problems, too (Corcoran and Wallich, 1992).

- Stockholders in these high technology companies, especially institutional investors, were on the lookout for and wanted opportunities for short-term gains.
- Timorous company managements were reluctant to put more money, especially their own money, into investment in long-term research and development.
- Company directors were also alarmed at the cost and difficulty of getting venture capital to support large in-house research facilities.

An obvious solution would be to outsource some research requirements to cash-strapped universities. They would be aided by government monies and tax concessions. At first glance, this would seem to be an excellent arrangement and no doubt in many circumstances it was.

Advantages or Disadvantages?

Such alliances should serve, so the reasoning went, many mutually beneficial purposes:

- First, the academic research worker had an extra income with which to supplement a university salary. This perquisite encouraged a valued and skilled academic to stay in an academic milieu.
- Second, the academic research worker provided industry with skilled professional assistance. At the same time, by adapting to the vicissitudes of the industrial life, the academic obtained broader experience with which to enrich the university, all while remaining within the cloistered halls of academia. The research worker, therefore, became a better teacher for graduate students.
- Third, the consulting academic was in a much better position to determine what were the most demanding problems facing industry. Then that academic could formulate a proposal of work that

208 ■ Food, Consumers, and the Food Industry

- Fourth, such liaisons allowed an exchange of facilities. Industry loaned its facilities to academic research workers to train students, to undertake tests with specialised equipment prohibitively expensive for the university, and to conduct large-scale production trials.
- Finally, graduates who had practical industrial experience often found employment with sponsoring companies.

It would appear that there could be no downside to what appeared to be a win–win situation.

It seemed plausible, so the universities thought, that science should be directed to support industry. Therefore, universities ought to identify industry needs and provide results useful to industry. To this end, they created Offices of Technology Transfer (OTT). Staff in OTTs canvass the resources of the university and the industrial community (whether local, regional, national, or international) seeking synergies with research or industrial interests. They review the assets of the university (research results and skilled staff) and determine if they are potential business benefits for industry. Patents and royalties can bring monies to the universities and gains to interested companies.

OTTs seek out industries willing to outsource research to their universities. They design training courses with the cooperation of university staff for companies to use to upgrade staff for companies.

Whether the mission of universities "…to be a haven for research and intellectual nurture of emerging concepts" would be jeopardised by closer ties to industry was questioned by Danzig (1987). He asked rhetorically whether research would dominate at the expense of education (another mission of universities) or whether "human effectiveness" would be lost in writing proposals, grant applications, and review systems for potential sponsors. However, his answers paid little attention to the potential for venality that can be tied to corporate donorships.

Many, including some academics themselves, challenge whether it is the function of the university through this dedication to industrial applications to turn out trained technologists. Nobel laureate, John Polanyi, speaking at a symposium held under the auspices of the Royal Society of Canada decried the policies of all levels of government to turn universities into "outlying branches of industry." The losses for industry and universities are "…breadth of knowledge, far-time horizons and (an) independent voice" (May, 1999). For Polanyi, research must be free from

of the Gairdner Foundation International and Wightman Awards is quoted as saying at the Minds That Matter conference (Arnold, 1999b). A further warning of his was "If professional values fail, professional prostitution results."

"There can be serious conflicts of interest when universities serve industry: Who pays? Who gets the credit? Whose interests conflict?" (Lane, 1996). Lane argued very forcefully that universities should measure their net accomplishments in "ideas and initiatives." At present, it would appear that universities measure the value of their staff in the amount of research monies they fetch. Universities also require intellectual pursuits (broadly, the arts, humanities, and theoretical mathematics) which are not germane to private industry but should be supported equally. These pursuits attempt to explain the human experience as science attempts to explain the material world.

More Concerns

Money always comes with strings attached to it. To receive money donees are expected to contribute something that the donor wants in return. Private industry, the donor, wants products developed that it can promote and sell. Private industry is driven by marketing departments. It is not driven by any altruistic thirst for knowledge. Consequently, such industry funding changes the science from pure science to applications technology, which this applied science has now become. Marketing departments have set the goals for science.

Goals-driven technology greatly influences scientists in how research is directed and in what questions will be asked. Therefore, the need to obtain private industry funding has direct consequences for:

- What assumptions researchers will formulate in their hypotheses
- How they will direct their thinking toward the goals of their donors
- How they will develop their experimental plan for their donors
- Ultimately what kind of research they do, and
- The selection and interpretation of the final results for presentation to their donors and to the scientific public.

Researchers would be very mindful of their donor's interests, very aware of where their funds for graduate students, laboratory assistants,

> *"All men are tempted. There is no man that lives that cannot be broken, provided it is the right temptation, put in the right spot."*

> Henry Ward Beecher in *Proverbs from a Plymouth Pulpit*

There are other concerns for universities in this alliance with industry. Private industry donors do make demands before they commit monies to any project.

- They may require approval of the experimental design in the proposal that a researcher submits before experiments begin.
- Donors may demand a prepublication review of any articles submitted for publication. Papers for which they have contributed monies, equipment, support for travel to conferences or experimental materials may be extensively edited prior to publication.
- Some sponsors may retain a right to refuse publication of any part of work for which they have made a contribution.

An example suffices to illustrate the points above (this issue, still unresolved, broke in the newspapers (*The National Post, The Globe and Mail*) in the winter of 1998).

> *Dr. Nancy Olivieri, a blood specialist known world-wide and head of the hemoglobinopathy program at the Hospital for Sick Children (Toronto), was studying the effect of a drug to treat thalassemia. She was the lead researcher in an international study using the drug. She informed the sponsor that use of the drug was potentially harmful and that she would publish her negative findings. In addition she informed the Hospital for Sick Children. The sponsor's reaction was swift. It removed Olivieri as project leader and threatened legal action. The Hospital removed Olivieri from her post. Only the University Faculty Association and the University's president stood up for Olivieri.*

Here was an issue of academic freedom. Experts can and do disagree and these opposing opinions should be aired in the proper forum. But certainly not aired in the courts as was threatened nor should what appeared to be a form of whistle-blowing be a cause for dismissal.

Was the carrot of large research grants that the sponsor could give to either the hospital or to the university an influence in their behaviour? Was

Noble or Naïve Behaviour?

There have been examples of questionable behaviour by researchers which does cause sceptics of science in the general public to raise their eyebrows (Day, 1998a). These occurrences have received prominent display in the print media. One example described by Day concerned researchers whose travel expenses from Europe to a closed conference in the U.S. were paid by a pharmaceutical company. These researchers had authored an article favourable to their sponsor's product, a product over which there was controversy surrounding its use in the research community. The authors' favourable observations may or may not have been influenced by the travel support they received. But the origin of any support either financial or material they received should have been made very clear to everyone reading the article.

Stelfox et al. (1998) reviewed English language articles published from March 1995 to September 1996 which described the use of controversial drugs known as calcium-channel antagonists. The startling results were that articles favourable to these drugs were nearly three times more likely to come from researchers supported by the drugs' manufacturers than articles written by researchers whose findings were either critical or neutral and had not received support. Most of the researchers who had been supported by the drug manufacturer did not indicate the source of their support. Only 2 of the 70 articles comprising the study gave any indication of a possible conflict of interest.

Stelfox et al.'s findings on the results of calcium-channel antagonists were reported in at least one Canadian national newspaper, one local (Montreal and environs) newspaper, an internationally distributed science magazine (*New Scientist*; see Day, 1998a) and one business magazine. How many other publications picked it up on news wires is anyone's guess. Certainly a public that included the lay public, the business community, the scientific communities, students, and general science buffs all of whom are customers and consumers, read the article and concluded that the researchers knew "which side their bread was buttered on."

Gori and Luik (1999) co-authored a report that strongly condemned the U.S.'s Environmental Protection Agency's work in finding secondhand smoke a human carcinogen. In their opinion, there was no reason to ban smoking in the workplace since secondary smoke was not a hazard in the indoor environment. How is the public to react when it was revealed that Luik had received research funds from the tobacco industry and Gori

high salt intake and hypertension. There has been controversy about this salt/hypertension link. The journal did not report that the signer of the editorial was a consultant to the Salt Institute (Day, 1998b). Acknowledgment of this financial connection is important as an aid to readers in evaluating the opinion expressed. More of the general public will see the article in the popular science magazine *New Scientist* than in the more esoteric journal, *Science*, and make their own conclusions about possible biased opinions.

How objective are scientists when conducting research on their sponsors' products? No bias may have been present in their research efforts, but scientists must ensure that there be no bias nor any appearance of bias in the findings they publish. The saying "Caesar's wife must be above suspicion" could be rewritten as "Science's exponents must be above suspicion."

Food scientists should not take any solace that none of the foregoing involved food research. The issue is that they did involve *scientists*. The incidents all received prominent play in the news media. It should not be forgotten that pharmaceutical companies are chemical manufacturers are ingredient manufacturers are biochemical companies and are thus companies with vested interests in genetically modified foods, in agrichemicals (chemical fertilisers, pesticides, and herbicides), and nutriceutical ingredients.

The results of science conducted by industry or by any vested interest group and the results and opinions of scientists supported by industry or vested interest groups must *always* be suspect. Scientists supported or employed by, for example, a manufacturer of a product:

■ On which many thousands indeed millions of dollars may have been spent in development and consumer research
■ Which has tremendous potential commercial value
■ For which there is a need to seek government or public approval of its use
■ For which there is some controversy respecting its safety and continued use

are subject to great temptation. They cannot be expected to be unbiased respecting any opinions on its use. The same pressures are on scientists conducting research for vested interest groups. At the very least, the

Researchers have great autonomy and discretionary powers in how they write papers, interpret data, and present their findings even in peer-reviewed journals. They can discard bad data points that they believe to be flawed or questionable; they can also select data and references supportive of their theories. They can omit data and references that are neutral or non-supportive of their assumptions.

Literature written by, and quoted opinions of, scientists who have been supported materially or financially by private interests ought to be interpreted with caution and scepticism. Financial support of research at universities by private interests has no other purpose than to serve the ends of those private interests. This support should be made very clear in any publication of the findings, conclusions, or opinions expressed; the authors may have a conflict of interest.

The value that can be had, and is needed, by a liaison between researchers in universities and corporate donors cannot be allowed to be reduced by marketing-directed corporate domination of academic research. The questions that are to be resolved are these:

- Where is this imaginary line in the sand to be drawn?
- How is the balance between industrial need for research and academic freedom to be reached?

The Role of the Media

The role of the media is to inform the public of the newsworthy issues of the day. Science is newsworthy. There are two elements that cause a distortion or blurring of the news of scientific issues, namely, the scientists themselves (it makes little difference whether they are university- or industry-based scientists) and the industry of journalism and its practitioners, the journalists. Both need one another. The industrial scientist wants public recognition that leads to a bonus. The academic researcher also wants public recognition and wants to whet the interest of potential investors or donors for further research. Journalists, especially investigative, scientific journalists, want *a story*.

Science and the Media

The penchant for scientists to call a press conference when they have

cause themselves and the departments and universities they represent some embarrassment. Sponsors want something developed that they can sell, promote, or advertise. The academic's normal reticence and usually careful language are not those of the marketing departments of the companies who sponsored their research. Here, marketing hyperbole will take over and academics can see their carefully worded conclusions blown out of proportion with marketing jargon. When scientists see the press releases from these marketing departments and see how their words have been interpreted they can very well be surprised and dismayed.

The industrial scientist is more understanding and tolerant of this exuberance on the part of the marketing department. In industry, the research and development section is often under the wing of marketing or its activities are closely coordinated with those of the marketing department.

The Journalists

There is a skilled body of technical writers and journalists that is growing larger every year as society becomes more technologically minded:

- These journalists are often technically trained individuals.
- They are skilled writers who write columns for daily newspapers using as their sources technical and scientific journals.
- They attend (and tape) press conferences that industrial and academic scientists call to announce their discoveries and, incidentally, in which the scientists attempt to garner government and venture capital support.
- They interview the scientists and researchers who have published their findings and tape these interviews for use on radio talk shows or for use in news releases.
- Journalists attend scientific conferences and write up their findings and also report on the impromptu meetings as well as unguarded and indiscreet comments at these conferences including cocktail and dinner conversations.

Research findings good and bad, peer reviewed and not, get published in the popular press. So, too, do the sometimes heated arguments and conversations between scientists become available for the general public to read. This includes the confrontations between experts, the refutations

A trip through any magazine store will reveal to the science *aficionado* a wide range of science magazines ranging from peer-reviewed journals to magazines with popular science articles on specific topics to news magazines on science. Most business, news, and other popular magazines, have a science column. Television and radio have regular programs devoted to science topics in which they attempt to delve behind the scenes and often do investigative science documentaries on controversial topics such as cloning, global warming, or genetic modification of organisms.

Anything newsworthy or controversial in science or food is reported and headline writers, a breed apart from the science journalists, are well versed in writing headlines that grab their readers' attention. Stories about clashes between scientists and their sponsors can be very dramatic, confrontational, and public. They garner headlines in daily newspapers, opinions in editorial columns, and wide television news coverage. Science is big news and good entertainment.

Controversy gets the attention of the public. The sight of two experts feuding on a split-screen on television does make for good journalism and good ratings. For example, Table 7.2 details the events of a week of sensational and highly entertaining reporting on food (Corcoran, 1999) by the Canadian Broadcasting Corporation (CBC) which occasioned the following editorial headline, "Hard to follow: Is CBC news safe?" in a Canadian national newspaper.

How do incidents like these influence the opinion of general public concerning science? They do receive broad coverage in all media including the Internet.

The appearance of reports of scientific misbehaviour in the press does little to improve the image of science and scientists in the public mind. The exhortation that something has been "scientifically established" is falling more and more on the deaf ears of the public.

To further find that experts may possess a certain amount of venality simply puts another nail in the coffin of trust the public may have had for scientists.

Summary

From Palin's plaintive "All I ask of food is that it doesn't harm me" which opened this chapter, discussion on food safety has run the gamut from

216 ■ Food, Consumers, and the Food Industry

Table 7.2 A Week of Questionable Reporting of Food-Related News by the CBC (Corcoran, 1999)

October, 1999	Episode
Tuesday, 12th	Radio News reported on a farmer who accidentally sprayed a toxic herbicide on his cows. Report likened incident to similar incident in Belgium which resulted in major agricultural losses and food recalls. [There was no problem in Canada; problem confined entirely to farm.]
	Food series entitled "Hard to Swallow" launched on nationally broadcast program *This Morning*. First installment devoted to a what-if scenario of hypothetical incidents like the above.
Wednesday, 13th	*World Report* reports that food system is unsafe because there are fewer inspectors.
	Second installment of "Hard to Swallow" highlights dangers to food safety in feedlots and slaughterhouses from killer pathogens.
	Radio documentary on food inspection system describes concern of veterinarians and inspectors over possibilities that diseased poultry meat may enter food system. [Labour dispute by inspectors' union not mentioned in documentary.]
Thursday, 14th	Radio news reported that federal government is suggesting a new risk assessment program. Federal veterinarians are quoted as saying companies "will be allowed to market unhealthy-looking poultry meat."
	This Morning in a continuing series interviews a prominent activist who states that Canadian beef is full of cancer-causing hormones and therefore banned in Europe.
Friday, 15th	Radio news reports on a Canada Health Coalition, members of which are ex-Health Canada Bureaucrats and activists, public forum in Ottawa. Attendees include head of Consumer Policy Institute in New York. Item suggests much of Canadian food has "potentially harmful growth hormones, antibiotic residues and foreign genes."

Table 7.2 (Continued) A Week of Questionable Reporting of Food-Related News by the CBC (Corcoran, 1999)

October, 1999	Episode
Saturday, 16th	Genetic Engineering, Pesticides and Labour in a Globalized Food System, a one-day conference sponsored by McMaster University's Ontario Public Interest Research Group. One of main speakers is author of the book *Farmageddon*.
Sunday, 17th	D. Suzuki, host of TV nature show, environmentalist/activist, and journalist/author, held a news conference in CBC headquarters in Toronto where he claims Canadians are being used as guinea pigs for genetically modified foods.
Monday, 18th	Canadian Health Coalition released a petition signed by some Health Canada employees warning of coming food disaster. They oppose legislation to change the food safety system.
	Health Minister promises to protect the integrity of the present system.

health has now often become a matter of conflicting opinion among scientists.

Science and scientists have been shown to prostitute their talents for financial rewards. Bias in science has been demonstrated. Scientists have become arrogant in assuming they are the only conduit for TRUTH. They disdain other opinions and conduits as irrational. Customers/consumers are irate that their voices are not respected in risk assessment.

Agrifood companies have not been open about their programs and intentions. Food manufacturers are quick to capitalise on the fears and hopes of their customers/consumers with new food products hyped to promote good health. Sleek advertising hyperbole blurs the consumers' already poor understanding of healthy eating.

Customers and consumers of food products are ignorant of food, its preparation and nutritional value. They cannot judge clearly what is safe and what is not, or what is nutritionally good or what is nutritionally unsound. The information available to them is confusing. They do not trust the experts and their opinions of what is or is not safe.

have ulterior motives for what scientific findings they communicate and how they do it. Vested interest groups with their own scientists and their public relations specialists further confuse the issues for the public by producing contrary expert opinion. The public's lack of basic education in scientific, medical, nutritional, or biological issues does not help.

The broad issues of the safety of foods and the communication of these issues rationally to consumers are complex:

■ Science must regain its respectability as unbiased and lacking venality and become a believable element in resolving safety issues.
■ Scientists must be more accommodating to opinions based on social, cultural, and ethical truths.
■ Governments must assume responsibility for the education of its citizens to become better consumers of science.
■ Journalism must develop higher standards for critical reporting of science in the various media.

Unless there is better resolution of these issues than there has been, conflict and confrontation respecting the broad understanding of the safety of foods and the ability of governments to rationally legislate it can be expected.

Chapter 8

Legislative Dilemmas in the New Millennium

"If the law supposes that," said Mr. Bumble…, "the law is a ass, a idiot."

Charles Dickens, *Oliver Twist*

"Here Lay the Dilemma"

Martin Luis Guzmán, *The Eagle and the Serpent*

A dilemma is defined as any situation in which a choice must be made between unpleasant and disagreeable alternatives. Legislators always face unpleasant and disagreeable alternatives as they attempt to promulgate sensible laws that regulate food and its commerce from field and sea to customers. One cannot find a sound argument against the *need* for legislation to regulate the various facets of food production and processing, its trade and food products themselves. There may, however, be sound arguments about its administration, application, and sometimes seemingly unnecessary complexity.

The dilemma arises directly from the tasks that the legislators must confront and resolve. Among these tasks are

1. Legislators must establish policies that provide an abundant, safe

3. They must support directly and indirectly the primary producers and harvesters of food to assure an abundant and continuing supply of food.
4. They must protect their manufacturing industries. In many countries the food segment of the economy is a major component employing a significantly large work force.
5. Legislators must protect the independence of the national food economy from foreign domination yet maintain international trade relations with other countries for safety of the food supply and to earn foreign monies.

These tasks, some of which appear to be contradictions, introduce unpleasant alternatives.

For example, to accomplish tasks (1) and (3) above introduces a cost factor. Food costs must rise. Customers who are voters might balk at increases they consider too large. Or food must be subsidized which necessitates taxation. Subsidies might contravene (5) above, trade agreements with other countries. Subsidies and other support programs will increase costs at the processing and retailing levels (4). Rising taxes and rising food prices will irritate the public.

Food, for all the above reasons, will always be a highly regulated item. Free trade in food, especially a global free trade, is very unlikely in the new millennium. The political, social, nutritional, and historical importance of food throughout the new millennium must be factored into any deliberations affecting food legislation and its attendant regulations.

To understand the dilemma and suggest avenues of resolution one must first examine the policy making and legislative processes as these exist commonly in many countries. With this appreciation, one can understand how food legislation can be, and is, manipulated for good or bad purposes. The limitations of the legal process in regulating food and its commerce either internationally or nationally are then clearly revealed. Such an analysis will assist one in understanding the public's and the food industry's often open dissatisfaction with the process and the active antagonism of many consumer advocacy groups toward the effectiveness of the legal process.

How the Dilemma for Food Legislators Arose

In short, food is government and government is food.

That the state would sell grain at a fixed price, especially in disastrous harvest years, is not a new concept. The production and distribution of grain were controlled about 2000 B.C. in China by the Emperor Shun (Spitz, 1979). Public treasuries maintained a stable price for farmers when grain was abundant and cheap; they bought up surplus grain. Farmers did not suffer. When grain was scarce and expensive, public treasuries released their surplus stored grain, maintained a stable price, and the people did not suffer because of rising food prices.

But there are other reasons for the government's intervention in food besides assuring that farmers are happy or that the urban population is kept fed and happy. Nobody became restive.

Food safety and quality as well as fair trade in food must be regulated. There are several distinct levels of governance that target safety, quality, and wholesomeness of food and food products, that define standards of identity, and that regulate its trade and commerce:

- International level. Agreements between one nation and another or amongst groups of nations will regulate trade. The Codex Alimentarius is an international code of food regulations agreed upon by many nations for use in such alliances. The World Trade Organization acts as a governing body for trade disputes.
- Levels of governance within nations. Most nations have a governing body or department (often several) which is responsible for the safety of food, its labeling and standards of identification, e.g., Food and Drug Directorate (Canada), Food and Drug Agency (U.S.), Ministry of Agriculture, Fisheries and Food (U.K.). Other government departments regulate agriculture, trade and commerce, health and welfare of citizens, and consumer affairs. At lower legislative levels, there can be provincial, state, municipal, county, or parish governments that promulgate regulations governing food, its trade, and agriculture according to the needs of the regions they represent.
- Nongovernmental level. At this level are organizations, some of which may have a quasi-legislative power, that are permitted to regulate the supply and flow of certain commodities or of some trade practices. Marketing boards, professional associations, and trade associations are examples.

Policies developed at these different levels ultimately result in various bodies of legislation (codices). These regulations or codes of practice are the means by which legislators hope the food supply and all its associated commercial activities can be safely and sensibly controlled.

How Food Legislation is Developed and Influenced

Food legislation is not new. It was needed very early to counter mischief by processors and retailers alike. The ancient code of Hammurabi (lived 2123–2081 B.C.) contained legislation regulating food standards and trade. Accum's analytical work (see Table 1.3) exposed the prevalence of gross adulteration of food offered to the public (Accum's analytical work was published in 1820). Such revelations led ultimately to legislation regulating food in the U.K. It also earned Accum the enmity of many English food manufacturers (Farrer, 1996).

Figure 8.1 presents a generalized overview of a typical legislative policy-making system. Shown in the upper portion of this figure are those groups that influence the policy-making process by either lobbying representatives or by presenting briefs at hearings called by policy makers. Legislation and regulations stemming from the legislation bear the imprint of these lobbyists and groups.

Non-Governmental Organizations

Marketing boards are commodity-based associations that are empowered to regulate the supply of food commodities (eggs, milk, poultry, wheat are usual targets for regulation). Their regulatory powers extend not only to the supply of the commodity, and hence its price to buyers, but they can even regulate the supplier. Through supply licensing arrangements given to processors, boards control who further processes the commodities they control, i.e., they can decide, for example, who obtains a license for market milk or for industrial milk destined for cheese manufacture or for other manufacturing purposes. They influence governments to impose quotas for foreign imports of the commodities or products containing the commodities they control.

The net effect of these boards is then to:

Permit them to manage the supply of commodities that they control

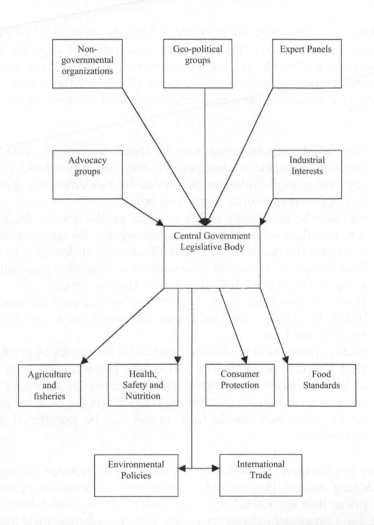

Figure 8.1   A Typical Policy-Making System with the Elements Influencing It and the Areas Influenced by It

Marketing boards practice supply management. At worst this practice is price-fixing. In any other business this practice is illegal. Governments impose heavy fines for price fixing in other areas of commerce.

Manufacturers pay an inflated price for raw materials and customers pay an inflated price for finished foodstuffs. For example, the Canadian

products are, of necessity, more costly in Canada. In similar fashion, egg marketing boards control the price and availability of eggs for further processing (see Chapter 2). The Japanese rice growers have protected their internal markets from foreign imports of rice with a similar effect.

There is a real economic impact resulting from the activity of marketing boards.

- First, food manufacturing jobs in industries where added-value products are made are lost because the cost of raw produce is too high for manufacturing to be profitable. Governments, therefore, lose a potential source of tax revenue.
- Second, the necessarily higher costs of production for these products with their resultant higher purchase price discourage customers from purchasing them. The extra financial burden due to higher food prices on Canadian customers, for example, was estimated as approximately $2 billion per year (Bryan, 1993).
- Third, the producer must support the hierarchy of the marketing board. Its activities and its employees represent a cost that must be accounted for.
- Finally, production quotas can lead to inefficiencies in production. They support the inefficient agricultural producer. For a producer to become more competitive and productive by applying newer technologies requires that the production quotas of other producers be purchased. Production quotas can be purchased at very high prices.

Many nongovernment associations and advocacy groups are opposed to marketing boards. For example, consumers' associations oppose the higher prices that are forced upon customers. Trade associations representing food processors also oppose the higher costs that their members must pay for their raw produce.

In defense of supply management controls, nations need to protect the independence of their food supply and maintain its independence from both foreign and from corporate domination. The primary producer must be protected by fair prices for commodities, from the dumping of produce below cost by foreign nations, and must be provided with security at a reasonable cost.

Farmers have a strong voice as voters and their elected representatives

Advocacy Groups

Advocacy groups by their very nature polarize people, for example,

- Urban associations vs. rural associations, and, hence, the polarization of elected representatives of political parties representing the rural vote vs. representatives of the city vote
- Groups representing residential interests vs. those espousing commercial and industrial interests whereby municipal governments are drawn into controversy over zoning bylaw changes
- Animal rights activists opposed to supporters of factory farming, or against animal experimentation, or against retailers displaying factory-raised meat or even against meat eaters themselves
- Consumer groups against farm marketing boards, against price support programs, and supply management systems

The strongest weapon of consumer advocacy groups is their ability to mobilize segments of the population (women's groups, seniors, parents, handicapped people, church groups, indeed, any marginalized people) to some form of demonstrative action such as sit-ins in government offices, picketing offices or plant sites of those manufacturers who offend them, or letter-writing campaigns to elected government representatives. Radical groups stoop to more violent actions that border on terrorism. They deliberately contaminate food or threaten with their intentions to contaminate food to publicize their cause.

Those groups that have most influence with food and its legislative regulation are

- Environmentalists
- Animal rights activists
- Religious fundamentalists

Environmentalists

Environmental advocacy groups have been influential in many areas of food production. They have opposed the conversion of recreational lands to grasslands or farmlands for food usage. They have blocked particular agricultural and animal husbandry practices carried out close to suburban residential areas and challenged existing farming practices near areas slated

Environmentalists have challenged the siting of fish farms, the clearing of forested areas for agricultural purposes, the applications of herbicides, pesticides, and fertilizers. They rail against the heady aroma of green manure applied to fields near suburban residential subdivisions yet espouse natural, organic farming.

The more militant of these advocacy groups have ripped up genetically modified crops wherever they were be found growing. They have pressured governments for labelling of food products containing, or made with ingredients derived from, genetically modified crops. They are very skilled at making their opinions heard and have mounted smart, provocative advertising campaigns against whatever they are opposed to.

Lobbying efforts of environmentalists resulted in the imposition of recycling practices which set in motion, like a domino effect, a series of events:

- First, a recycling industry was born. This necessitated transportation for the pick-up and sorting of recyclable materials and distribution to recycling sites.
- New industries were spawned for sorting waste, de-inking paper, and creating products from the recycled material.
- Retailers were saddled with container recycling depots within their stores. They were required to set up and institute a refund program.
- Many countries passed legislation requiring that all packaging materials must be manufactured using only materials which can be recycled. This has led to research into new, often more costly, biodegradable materials for food packaging.

There is a growing concern about the wisdom of some recycling practices.

- Are some recycling practices themselves the source of even greater pollution? Certainly de-inking operations do create another pollutant to be safely disposed of.
- Are recycling operations energy efficient? Some have suggested they are not. If all energy consumption from pick-up to final distribution to plants using the recycled material is considered, some energy recycling models suggest an energy imbalance with added pollution.
- Do they present an added layer of hidden costs to foods? They

Furthermore, while many in the public espouse recycling and want recycling plants, they do not want them situated "in their neighbourhood."

Animal Rights Activists

Animal rights activists object to any practices that might involve the mistreatment of animals. They are consequently against the intensive factory-farm raising of animals. Many are vegetarian and object to the use of animals as food.

Their activities are largely poorly thought-through acts of vandalism. They have been known to spraypaint packages of meat in meat counters or inject (or threaten to have injected) turkeys with unspecified poisons. Leather clothing wore by people has also been defaced with spray painting.

Religious Concerns

There are groups who believe that social, ethical, and religious concerns should be important elements in forming legislation. For example, Arabs and Jews alike want to see religious standards as factors in legislative considerations. They want to have the option to condemn food imports on religious grounds, that is, to ban foods which do not conform to certain religious requirements in its preparation or slaughter.

Others would like to have provisions in the law that restrict the movement and importation of food products manufactured by socially, ethically, and environmentally irresponsible companies or countries.

Geopolitical Groups

This group includes most if not all political parties. All have in common closely defined economies based on their geography, for example, in North America:

- The Prairie Provinces in Canada or the midwestern states in the U.S. where cereal crops and livestock production predominate, the rural lifestyle predominates.
- The Maritime regions where the fisheries industry is a key economy.

Representatives are appealed to by their constituents, by local manu-facturers, and organizations which have provided financial, material, and volunteer support, who each have their own agendas. Such partisan activities influence food legislation.

Any political group whether it leans to the right or the left in its philosophy, whether it espouses environmental causes or not, will put its stamp on any legislation it enacts. In parliamentary processes each political party will manoeuvre to alter, modify, or introduce legislation to advance its cause or beliefs. Problems are confounded when individual legislators have, in addition, their own hidden agendas concerning the interests of their local economies.

Expert Panels

Expert panels are called in by governments for informed opinions on unique topics of learning. For issues concerned with food and nutrition, these panels consist of prominent food scientists, nutritionists and dieti-tians, biochemists, agronomists, and consumerists, indeed all those asso-ciated with food, its production, its manufacture, and its consumption.

From the premise that all science is logical and objective, governments hope to receive a rational, scientific basis for any legislation that treats technical issues concerning food. Since elected representatives are often not scientifically trained, they are understandably confused by conflicting scientific opinions and by scientists feuding over interpretation of data.

Governments, in these situations of conflicting expert opinion, have two options:

- Hold any legislation in abeyance until a clear opinion is available
- Pass legislation based on the best available information

This latter action can be damaging. First, it can harm any political party that enacted the legislation should their legislation prove wrong. Second, legislation, once in place, whether it be right or wrong, is very difficult to change, c.f., the Delaney clause.

It is the interpretation of what is the "best available information" that brings out the best in the lobbyists and other vested interest groups. Politicians do not always proceed as expert panels advise on food issues. The reason is simply their need to make compromises with the many

Where Food Legislation Reaches: The Long Arm of the Law

The bottom half of Figure 8.1 demonstrates in general terms where food legislation reaches and Table 8.1 depicts those areas where the groups in the top half of the figure might have their greatest impact in influencing legislation.

By far, the food microcosm of any nation is the most heavily legislated or very close to being the most heavily legislated segment of industry. All international, national, and even nongovernmental legislative levels have some regulatory influence on agriculture, fishing, and food processing in general as well as a far-reaching impact on commercial activities associated with food and its economic importance.

The Quest for A Safe and Wholesome Food Supply

"Edible, adj., good to eat, and wholesome to digest, as a worm to a toad, a toad to a snake, a snake to a pig, a pig to a man, and a man to a worm."

Ambrose Bierce, *The Devil's Dictionary*

Legislation, if enforced, does provide some modicum of assurance of a safe and wholesome food supply (Table 8.1). It also provides some protection from fraud for the customer and the consumer.

By enforcing food regulations through inspections and analyses and prosecuting where offences are found, food sold within a country should be safe. Or is it? Countries with strict food legislation or with a long history of such legislation or employing strict food regulatory and enforcement agencies do not necessarily have food supplies that are free from problems of public health significance. No amount of food legislation, no matter how strongly it is written nor how vigorously its enforcement is carried out, can guarantee either the quality or the safety of all food consumed within its borders.

For example, in the U.K., in 1997 more Britons suffered food poisoning than had been recorded since records were kept (Coghlan, 1998a). Britain has an excellent library of food legislation. The Agriculture Department of the U.S. in April, 1999 ordered a recall of meat products including hot dogs, luncheon meats, and various sausages made by an Arkansas com-

Table 8.1 Overview of the Extensive Reach of Legislation Pertaining to Foodstuff

Area of Impact	Elements of Food Microcosm Regulated
Agriculture and fisheries	• Siting of farms and fish corrals • Use of pesticides, herbicides, fertilizers permitted • Antibiotics and pharmaceuticals to be used for animals • Quotas for fishing and commodities • Farm and fisheries assistance programs
Processing	• Site location for plants • Design and construction materials for plants • Adherence to processing codes of good manufacturing practice for safe handling, manufacture, and storage of food • Worker safety standards • Import, export permits
Product	• Commodity grades • Standards of identity and grades for some basic products • Lists of approved additives (or lists of restricted additives) for food use • Establishment of limits for the presence of toxic or proscribed chemicals or agricultural residues in foods • Standards for extraneous matter in foods, limits for microbiological hazards
Package	• Package sizes • Composition of packaging materials in contact with food • Product net weight and contents regulations • Product nomenclature and product information • Label requirements for ingredients and nutrient content
Marketing	• Advertising claims respecting nutritional and health benefits • Promotional and advertising guidelines to prevent

Table 8.1 (Continued) Overview of the Extensive Reach of Legislation Pertaining to Foodstuff

Area of Impact	Elements of Food Microcosm Regulated
Retailing	• Zoning regulations for store locations • Retailing hours
Environment	• Recycling of packaging materials • Zoning bylaws • Odor and noise abatement programs at food plants • Waste water recycling at plant sites • Water projects for farm lands
International Trade	• Trade alliances and treaties • Tariffs and non-tariff trade barriers • Anti-dumping regulations

to food safety and inspection as well as one of the most comprehensive bodies of regulations and legislation pertaining to food in the world. Japan, at the end of the last millennium and into the summer of 2000, has suffered large outbreaks of food poisoning yet it, too, has stringent food regulations.

An active campaign of inspection of restaurants and the citation of offenders with their names and their offences published in the daily newspapers in Montreal have done nothing to alter either upward or downward the incidence of foodborne disease in the Montreal region (Idziak, 1998).

Food legislation can be developed only to answer to known hazards associated with foods. Knowledge of food hazards is based solely on those processes whose history, so to speak, is recorded respecting their value in food safety. Nevertheless, the process of making salami, a food with a long history, is being questioned regarding whether it is a safe process against new variants of an old microbiological hazard. Regulations cannot safeguard the public:

■ Against new hazards which may be associated with novel or innovative foods;
■ Against food processes with an unknown history of safety or which

This lack of history respecting the long-term safety of genetically modified foods is one argument that those opposed to them have used.

Legislation also extends to the premises in which food is prepared or stored. Building codes, processing regulations, or codes of practice are attempts at ensuring that the environment in which food is prepared is such that food is not contaminated during production or that the preparation or the processing is designed to assure a safe and wholesome product. They assure that the purveyors of foods (or their distributors) in the many different marketplaces receive a safe product. Customers and ultimately consumers should receive safe food products.

Quality

Food quality cannot be legislated directly. Legislation may provide:

- Grade standards for foods such as meats, fruits, and vegetables with each grade representing different characteristics of composition, texture or color; or
- Standards of identity for composite foods which establish that the food meets the minimum composition for what it is called.

This is not necessarily that elusive characteristic called "quality".

Quality can be managed within a plant only by combinations of processing characteristics involving raw materials, ingredients, process parameters, and stringent sanitation to maximize the desired quality feature. Quality must always be associated with respect to some characteristic, for example, quality with respect to color, or with respect to nutritive content, etc. Grade standards for products are merely minimal standards of composition, solids content, viscosity, particle size, and integrity, etc. at which producers and manufacturers alike aim. Few manufacturers would exceed the standards because their costs would increase over the costs of their competitors.

Labeling regulations verify to consumers that the food is what it is stated to be and that it meets the standards established for the grade or the named product. Label statements cannot be described as, nor are they intended to be, guarantees of quality. They describe only that the product within the container adheres to certain minimum characteristics specified for that product. In essence then, grade and identity standards are those

Self-Regulation

If then legislation cannot be used as an efficient tool to assure either the complete safety or the quality of the food supply, is their a need for such a volume of legislation?

There are some who believe that minimal regulation of the food industry would be much preferred and would serve the public better. Their solution is self-regulation. The argument is that

- Competitive pressure would force all food manufacturers to maintain high levels of quality and safety in the products they offer to the public; and
- Adverse publication of any public health concerns for a product would cause such economic fallout that a company could be destroyed.

In short, peer pressure and fear would drive safety and quality. If there were self-regulation, manufacturers would vie with one another to have the safest products with the lowest microbial counts, etc., for fear that if they did not they could face economic ruin.

This is preposterous. It is just as likely that manufacturers would be encouraged to cut corners to maintain a price advantage and that product quality would drop across a broad range of products to minimal standards. Self-regulation cannot be a substitute for government intervention through inspection and analysis:

- Not all food manufacturers can be trusted to maintain the minimum standards that are required by current law and, as the saying goes, the bad apple spoils the barrel for the rest.
- A long history of bad apples has confirmed the previous statement and has caused the proliferation of food legislation now in place worldwide.

My personal experience, covering nearly 30 years as a consultant serving largely in the areas of quality management, product development, and crisis containment, have developed my strong reservations toward self-regulation. Too many horror stories have been told to me by quality managers who were pressed by production managers to meet production quotas or by sales managers for product

The Power of Food Legislation as a Weapon in International Trade

In the earlier years of the past millennium fertile land and abundant water supplies were resources to be fought over. They meant food and ample food meant protection for the tribe, clan, or nation. During the Middle Ages, the European Guilds were less crude. They set standards and effectively banned from their trading blocs any goods that did not meet their standards, had not been processed according to their work practices, or did not conform to their pricing schedules.

Legislation regulating food from the producer to the consumer can serve effectively as economic weapons for countries by which they can:

■ Coerce multinational food processing companies to invest in the local economy. For example, if a multinational company requires a necessary raw commodity, e.g., cocoa beans, a supplier country may require that the company do some processing locally.
■ Ban the importation of, or impose high duties on, foreign products that compete with a similar local product.
■ Limit the importation of foreign products which do not contain some percentage of local content, i.e., labor, raw product, or ingredients.
■ Protect local food processors, fishers, and farmers against imports through subsidies or tax breaks.

There are other weapons. Non-tariff trade barriers are tools through which nations attempt to break, find loopholes in, or circumvent their food trade obligations and protect national interests. As such, non-tariff trade barriers can be useful tools to get around trade treaties. Such protection will take a variety of forms:

■ Contravention of a nationally accepted standard of identity. A product identified in one country is not so identified in another country. A national dish, for example, may be defined in one country by a list of permitted ingredients. No product of the same or similar name is permitted unless it is composed of only those ingredients, often in the same proportion, and made in the same manner and style.
■ Nomenclature. A named product in one country is not so-named in another. This most frequently occurs when a country's name is

- Presence of agricultural contaminants. By setting very low toler-ances for pesticide residues, or for trace metal contaminants (including radionuclides), or for filth and other extraneous matter, etc., nations have a means by which products can be banned from importation. Included here would be the use of growth hormones in cattle which is the cause of the European ban of North American beef.
- Presence of restricted processing aids. Additives such as emulsifiers, antioxidants, texturizers, gelling agents, etc. used and approved in one country may not be approved in the importing country and may be used by that country to restrict the importation of products.
- Establishment of microbiological levels for raw or semi-conserved produce. When raw produce comes from exporting nations that have questionable sanitary procedures in food handling, an impor-tation ban can be very wise.
- Conformance to some international standard. Many companies require that their trading partner/supplier conform to or be certified in some international standard such as the International Standards Organisation's (ISO) for quality control or that products conform to the Codex Alimentarius's standards for products or processing. They usually are not sufficient to be considered trade barriers but do deter trading between companies.
- Accusations of dumping practices by the importing country against the exporting country. The exporting country is shipping product into another country at below its cost.
- On the basis of claims of unethical or unfair business practices, violation of workers and unsafe working conditions, and/or human rights violations in the country of origin of the products. This was a contentious issue at the Seattle Round of the World Trade Organization (WTO).

Any of these could be used as a basis to prevent the entry of goods processed in one country from entering another country. The usual goal of such an action on the part of an importing nation is protection of a specific commodity-based industry or maintenance of prices through a form of supply control management.

Classic examples of open feuds that have occurred are

The banana wars between the U.S. and the European Economic

interests. The EEC preferred instead to get their bananas from their old colonial holdings which were under the control of EEC companies. The U.S. threatened retaliation with heavy tariffs on EEC goods. The dispute was eventually settled through the World Trade Organization.

The chocolate truffle wars in late 1998. This was a dispute between cocoa-producing countries and chocolate-manufacturing countries over the definition of chocolate. The cocoa-producers wanted chocolate made only with 100% cocoa butter to be called chocolate but the chocolate manufacturing nations wanted the addition of non-cocoa butter fats permitted.

The controversy brewing over the use of growth hormones in beef cattle in the last years of the old millennium. Europe banned imports of such treated beef on the grounds of safety despite rulings to the contrary by the World Trade Organization. Canada and the U.S., which have approved their use, have threatened retaliatory action by banning European food products.

In the last months of 1999, France and the U.K. were having a food fight. France refused to allow the importation of English beef fearing still the spread of "mad cow disease." U.K. shopkeepers and restauranteurs refused to sell French wines and cheeses or serve French products.

Such tactics would seem to contradict the concept of a global marketplace and put severe obstacles in the path of companies wishing to centralize operations for the sake of cost savings.

How these contretemps can be overcome is not at all clear. More importantly it is not at all clear that all nations want such issues resolved. They do serve a useful purpose.

At this stage in attempts at globalization of the food trade it would appear that to market globally food processors may be compelled to build locally and sell regionally what the local customers want.

Harmonization of Food Laws

A major task in the new millennium will be the harmonization of food laws worldwide. The foregoing section should be an indication that not all countries want harmonisation. Indeed, they may actively resist such a

- The codes would prevent the international trade of diseased, contaminated food or food otherwise unfit for human consumption and ultimately protect the lives of consumers.
- Food laws protect the reputation of nations and the livelihood of their farmers by guaranteeing that only quality foods would be put on sale internationally.
- They would protect food exporters from improper imposition of spurious health or safety standards as tariff barriers by importing nations.

Previous attempts at cooperation, except for the FAO/WHO Codex Alimentarius Commission, have been in the main unsuccessful. The Codex Alimentarius Commission has harmonized food regulations by developing a set, that is, a codex of food standards. It has had some success in international trade.

The history and workings of the Commission and the Codex Alimentarius were described by Adams (1983). The listing of standards that have been accepted and described in this reference is sadly out of date. The Codex itself is simply a set of standards for commodities, for fats and oils, and procedural rules dealing with more general topics such as hygiene, food additives, labeling, and presentation, pesticide residues, sampling, and methodology for analysis, etc. The rules are not imposed on members of the Commission but are meant to serve as guidelines only. They are entirely voluntary in their adoption.

Many nations that favor the use of the Codex Alimentarius have suggested that it be adopted officially in international trade. Nations opposed to this feel that its standards are too low, that is, below levels now in place in their own countries (the ugly head of non-tariff trade barriers?). Some countries want the inclusion of a right to reject food on the basis of religious or ethical grounds in the Codex Alimentarius. Such a move would put a purely scientific basis for these food standards side-by-side with a non-scientific, moral standard.

A Scientific Basis for Standardization: Whose Science?

Science as a basis for food standards such as the Codex Alimentarius raises many concerns (see Chapter 7). The question certainly must be posed: whose science? Everyone would agree that unbiased science would

Many governments require that chemical companies supply all the safety and efficacy data for additives that these companies wish approved for use in foods. This is not always unbiased science. Even if government scientists review the data there will always be an element of doubt about data supplied by scientists who are beholden to the chemical companies making the submission. Sponsored science may not always be reliable, unbiased science.

The same concern can be expressed at the international level. That is, data submitted by one country are equally suspect as they would be if submitted by an individual company. Hidden agendas may be being pursued by the nation submitting the data, for example:

> *While with the British Food Manufacturing Industries Research Association (later to become the Leatherhead Food Research Association), I served on an international committee whose mandate was to establish international standards for dehydrated onions. The experience was very revealing. Representatives from two different countries both of whom were large exporters of dehydrated onions almost came to blows because the two different sets of standards were designed so that the other country would have been excluded from international trade.*

The political machinations behind the adoption of the Codex in international trade have been described more fully by Walston (1992).

The only answer would seem to be the establishment of an international body of scientists or an international consortium of recognized scientific establishments either of which would be approved by all nations to monitor all submissions of food standards. This is basically how the Codex Commission now works. It has been my unfortunate impression while working with Canada's members on various subcommittees to the Commission that many countries employ lawyers to protect their countries' interests in Codex Commission committees, a practice that prolongs decision making.

Consumer Legislation

This has been discussed. Suffice it to say, there are three broad areas in which legislation is used to protect customers and consumers alike:

1   Weights and measures: Regulations have been drawn up to assure

2. Labeling regulations. Proper nomenclature, truthful product pic-
   tures and product descriptions, lists of ingredients in descending
   order of quantities, nutrition labeling, and the address of the
   manufacturer are all designed to protect and assist customers in
   making wise food choices.
3. Retailing fraud. Many countries or jurisdictions require that food
   price be prominently displayed on the package and on the store
   shelf. In addition, a unit price is required to be displayed on the
   shelf, i.e., $x per 100 g.

With the introduction of a new arena for selling food, a new hazard
has appeared. The use of credit cards, then the introduction of debit cards,
and finally the rapid growth of e-commerce have brought about a need
for consumer legislation on at least two more fronts: (1) security of the
transaction whether it be e-commerce or by plastic card and (2) privacy
of the customer making the transaction.

Security on the Internet is a major problem for the sale of all articles,
not just food, but e-commerce is still a fledgling industry. Fraud will grow.
Web sites can be hacked, broken into, and confidential information stolen.
For the food industry this has not yet emerged as an obstacle for food
retailers to address.

Security of credit and debit cards has been a problem almost from
their first introduction into commerce. Lost or stolen cards, unethical
merchants, carelessness of customers respecting the protection of their
passwords or the safe storage of their receipts are all contributing factors.
As more and more meals are eaten away from home, the use of cards
for payment of meals is becoming much more prevalent and much more
of a problem.

Privacy is a much greater concern for the general public. Consumer
advocacy groups are protesting to governments about the information that
can be obtained in today's commerce and that is sold without the card-
holder's permission to third parties for their use in marketing.

Environmental Legislation

The food industry is a major polluter and must be ranked up with many
of the heavy industries that are often cited for pollution. Pollution begins
on the farm with the use of chemical fertilizers, pesticides, and weed

The next pollution step takes place at the processing plant. (Transportation of the animals by road or rail to the processing plant should not be overlooked as a significant source of pollution and energy utilization.) Produce must be peeled, trimmed, and washed; animals need to dehided, defeathered, descaled, eviscerated, etc. Byproducts are accumulated and must be transported away or processed into useful products. Water, in large quantities, is required. This must be treated before it can be returned to rivers, streams, or sewage systems. Odors as well as noise are produced and must be controlled.

Food produces waste wherever it is prepared, served, and consumed. There are packaging materials, trimming waste, plate waste, used plastic cups, plates, and utensils, aluminium and glass containers to be disposed of. In short, the whole food chain from start to finish is a polluter.

Consequently, the result is laws:

- To enforce recycling or re-use
- Against littering
- To impose taxes on products whose packaging is not recyclable and extra fees for "too much" garbage
- To establish purity standards for waste water effluent, noise pollution levels, and to require odor and emission controls.

All of which probably makes, along with the other legislation discussed earlier, the food microcosm the most heavily legislated entity on earth.

The Dubious Result

But to what avail? Is the food supply any safer? Is the environment any cleaner? Even in developed countries the number of food-related illnesses continues to climb. Credit card fraud has been described as a growth industry.

Some experts claim that many anti-pollution restrictions and laws cause more pollution than the problems for which they were created to remedy. Many of the recycling programs as currently practiced only lead to more pollution.

The imposition of legislation requires that it be enforced; this requires a cadre of enforcers or inspectors — and tax money. For example: The city of Montreal like many cities has a problem with a lack of landfill

infrastructure to be built which must support an inspection system, which must have a means to monitor or measure the abuse, and which must have a means to litigate. It also requires that householders must shred sensitive papers, credit and debit card statements, and receipts which were formerly mixed in with wet garbage (not the best disposal technique but commonly done). This all costs money.

Food Legislation: A Summary

As shown in Figure 8.1, legislation pertaining to all aspects of food from production, processing, to its consumption cannot help but be influenced by partisan groups from all sides. Each political party and each vested interest group will attempt to alter, modulate, or introduce legislation to advance the particular political opinions or the causes they espouse.

Governments must protect their food industries, besides protecting the health and well-being of their people. The necessity for food safety and trade legislation, therefore, cannot be denied. This can only be done through legislation designed to protect agricultural and manufacturing interests. Such legislation, however, is not a guarantee of safety of the food supply or for fair and easy trade in food products between nations or within nations.

The solutions undertaken by many governments, therefore, have followed one of three paths:

1. Outsourcing duties of analyst and inspector to private enterprise, often to universities with food science departments or to designated qualified private analytical laboratories.
2. Spinning off government departments whose responsibilities it was to inspect and analyze food products thus they become private monopolies. These entities charge fees for providing these services to their clients, i.e., the food industry. That is, they become self-supporting, money-making entities.
3. Putting food companies on "scout's honor" to maintain and record a high level safety and quality controls in their manufacturing processes. That is, food companies who have shown that they have the resources and organization, are allowed to self-regulate according to a strict code of practice.

therefore, had to send water samples to private laboratories. The discovery of the contamination was duly reported, *so it was claimed*, to the proper department within the Walkerton municipality. The private laboratory felt under no obligation to inform anyone but its client. At the present time, it remains for a public inquiry to determine how the true course of events leading to the tragedy played out.

An interesting ethical question can be raised by the above. Does a private laboratory contracted to do confidential analyses for a client have an obligation to reveal to the proper authorities the results should they indicate a potential public health hazard? Are whistle-blowers to be encouraged? A perusal of the history of the more sensational stories of whistle-blowing incidents suggests that those who speak out are prone to being harassed and blacklisted by those they speak out against.

Whether, how, and at what cost food could be made safer are moot points. Many of the food poisoning episodes that are reported can often be traced to human error or ignorance in the preparation and handling of food or mismanagement of safety protocols. This lapse of safety in the food chain cannot be corrected by legislation.

Chapter 9

So Now What? The Issues, The Problems, Maybe Some Solutions

"To almost all men, the state of things under which they have been used to live seems to be the necessary state of things."

Thomas B. Macauley

"He that will not apply new remedies, must expect new evils."

Francis Bacon, Essay of Innovations

"It is better to know some of the questions than all of the answers."

James Thurber

Into the Maelstrom of the Third Millennium

"Tradition is a guide and not a jailer." With these words Chapter 1's review of the past thousand years of political, technological, and food develop-

History should open up horizons for new thought, not confine thinking to either maintaining the *status quo* or to continuing in linked step with past events.

Bacon's aphorism is much more forthright. There is a need to constantly evaluate new technologies to be able to confront future dilemmas. Traditional remedies are not to be rebuffed; there must be a willingness to accept evolutionary, even revolutionary, processes to resolve problems such as those facing the food microcosm.

The past can give a perspective for the issues to be faced in the future. It provides insight into:

- How technical and political events affecting the production, and marketing and selling of food arose to become issues of concern;
- How these problems were resolved; and
- How opportunities arising out of past events were used advantageously.

This understanding can be used as a framework around which to devise solutions to parallel situations in the present day. Wallace Stevens stated it succinctly: "All history is modern history." All is change but *plus ça change, plus c'est la même chose.*

The preceding chapters highlighted some major issues affecting the food microcosm that require resolution in the third millennium. Some issues have proven to be stumbling blocks to the acceptance of technical developments that could possibly benefit mankind, such as food irradiation or genetically modified foods.

This chapter will present the issues more directly and pose questions which are facing all members of the food production chain from primary producers to consumers. By providing awareness of the needs and issues and by asking questions perhaps some illumination and clarification can be provided. *How* they will be finally resolved cannot be foretold. Futurists, experts, and psychics, the latter many corporate leaders do seek out, all have a notoriously bad history for accuracy when judged by past records of their yearend predictions. It is left to others to formulate a viable *modus operandi*. As Thurber comments, it is better to know some of the questions beforehand than it necessarily is to know the answers to them.

relationship with technology. Ausubel (1996) presented two quarrels against the intrusion of technology into humankind's existence:

1. "Technology's success is self-defeating." That is, as technology provides more services for mankind, more problems are ultimately created in tandem with the progress.
2. Humanity does not always have sufficient wisdom to use new technology wisely.

For example, better public health and nutrition programs have increased life span with all the attendant problems associated with an aging population. The many advances in computers and electronic communications have speeded up the pace of business. People work more efficiently, accomplish more, are able to, and often must, make instantaneous decisions based on the real-time data they receive, and they can work at home. The result? People are suffering greater stress; work loads and the pressures that accompany them have increased. The home is no longer a restful haven from work stress. Workers feel less secure. The creation of environments and opportunities for reflection, contemplation, just plain thinking, and even relaxation are rapidly disintegrating.

"Men have become the tools of their tools."

Henry David Thoreau, *Walden* "Economy"

It is too early to judge Ausubel's second point, i.e., whether advances in food science and technology can fuel either good or evil intentions in humanity and whether humanity is wise enough to choose well. Advances in food technology have polarized views on contentious food topics (see also, Fennema, 2000); for example:

■ Genetically modified crops and animals
■ Factory farming (intensive farming)
■ Irradiation processing of foods
■ Environmental concerns arising from land use and water pollution.

Whether wise choices will be made on these issues remains to be seen.

engineering, i.e., through the rational approach. There must be an inevitable clash at some time in this third millennium between the technocrats and the non-technocrats. Thompson has described it thus:

> *"...it is the Technological Society that is totalitarian, for it reduces all cultures to mere ideological impediments to the advance of "rationalization." Technology is "total" because it sees everything other than itself as reactionary, irrational, and primitive" (Thompson, pages 164, 165, 1972).*

Mankind has always rebelled against totalitarianism.

Science can explain the material world but even here its explanation is imperfect. The causal relationships of the physical world become more comprehensible as previous partial and imperfect knowledge is discarded because it failed to explain adequately. More precise knowledge is used to build a new set of beliefs. But science's explanations of the material world can never become a certainty. To apply this rational approach of science to the human world with all its nuances and complexities is itself irrational.

Non-technocrats voiced strong opposition to policy makers, scientists, and other experts at the Seattle meetings of the World Trade Organization. Effective opposition has been waged against genetically modified foods — and all raised at a grassroots level. The public's growing distrust of science, and hence of any technocracy spearheaded by science, and some reasons for it were discussed at length in an earlier chapter.

Resolution of this chasm between scientists — of which group food scientists must be included — and the general public with their social, ethnic, cultural, and traditional trappings is necessary and important since this general public constitutes the food industry's customers and consumers.

Both scientists and customers/consumers (the public) must trust, accept and, if not accept, then respect the verities each contributes to resolving contentious food-related issues. Many of the benefits that are to be gained within the food microcosm will depend on such a rapprochement in the new millennium. If the populations of the world are to be not just adequately fed, but have available a wide variety of wholesome food of an acceptable quality at a price they are willing to pay to feed themselves, there must be a reconciliation acceptable to technocrat and non-technocrat alike.

The Legacy from the Second Millennium

The first chapter, an historical perspective, introduced some issues developing rapidly at the end of the last millennium:

- The world's population is growing. This growth, at some unforeseeable time in the third millennium, will be a burden upon the earth's resources of arable land, harvestable ocean assets, and supplies of clean fresh water. Unfortunately, none of these resources are increasing.
- Arable land is being encroached upon by urbanization, and destroyed by erosion, by pollution, and by desertification. As long ago as 1978, it was reported that cropland destruction was proceeding at a rate between 6 and 12 million hectares annually, a rate that was more rapid than the amount of land being added to the cropland base (Anon., 1978).
- Clean fresh water is becoming a limited and prized resource. Pollution of fresh water is increasing. Rainwater runoff from fields contaminated by pesticides and chemical fertilizers flows into streams and rivers. These, in their turn, pollute lakes and oceans that they feed into. Fresh water has already become a trade issue between Canada and the U.S.
- The oceans are being over-fished. Fish stocks are dwindling; governments are being forced to adopt conservation measures and have had to enforce these measures in their territorial waters with gun boats.
- The oceans are also being polluted by fish farming operations whose use of medicated fish food has spread beyond the ocean pens to contaminate wild stocks and to cause algal blooms and disease.

History, then, has confirmed one of Ausubel's (1996) points: Technology's success is self-defeating. Technology's success has contributed to world population growth by increasing humankind's life span, decreasing infant mortality, and lessening mortality from diseases.

There is presently drought in Africa and in the midwestern U.S. in the past two years while in India there has been by turns drought and terrible flooding at the start of the third millennium. The total amount of water in the world remains a constant but unfortunately, much of it is unavailable

Feeding the World

Feeding the world will be a continuing struggle between the number of mouths to feed and the amount of wholesome, safe food that can be harvested from the world's available arable lands and viable marine resources.

The question was put succinctly by Smil (1985) in a discussion of China's food problems but it applies equally well to the global challenge:

> *"How well can China, farming a mere fifteenth of the earth's arable land, feed more than a fifth of the world's population?"*

Chandler (1992) presents an even more pessimistic view. He points out that the world's food grain production is falling behind the rate of growth of the world's population. For him, the greatest threat to the welfare of man is the present rapid increase in population growth.

Many, on the other hand, naïvely believe that providing an adequate supply of food is the answer to feeding the world's growing population. They look to technology to resolve this problem and, in particular, pin their greatest hope on biotechnology and genetic modification of crops. They would produce:

- Crops that produce greater yields
- Crops with higher nutrient content, e.g., more high quality protein, so that available produce would be more nutritionally complete
- Crops with greater resistance to pests and plant diseases to prevent field losses and so increase yields
- Crops more tolerant of marginally arable land (tolerant of high saline conditions) and adverse growth conditions (cold or heat) to allow the use of presently non-productive land
- Technology to produce more energy efficient animals

These are very laudable avenues to produce more food. Certainly, they should be part of the armory of techniques but do they represent naïve, technocratic thinking? Yes, they do. All of them cost money, which most of the world's farmers do not have. The biotechnology complex wants payment for their vast expenditure of money on development.

predictions of Chandler (1992) regarding increases in grain supply lagging behind population growth.

An abundant supply of food alone does not eliminate starvation and malnutrition. The poor and needy in all countries, including those in developed countries, do not have enough food at the present time when the world does have an abundant food supply. How can they? The poor and the hungry must have access to this available food and they must have something to exchange for food (Streeten, 1983; Cracknell, 1985; Chandler, 1992).

Economic Problems

Swaminathan (1992) reports that the ratio of real per capita income between Europe and India and China was 2:1 in 1880; was 40:1 in 1965, and in 1992 was 70:1. This growing disparity drives trade imbalances which in turn spreads poverty among those countries with the most rapidly growing populations. This introduces a non-scientific, indeed irrational, twist to the problem.

Agricultural economic policies meant to encourage farmers with higher prices for their food conflict with the equally pressing need to provide low cost food to the poor. The two clash. Food pricing policies have been known to cause some undesirable side effects, e.g., the EEC's over abundance of wine, Zambia's food shortages, and the U.S.'s wheat glut (Cracknell, 1985). Legislators are caught up in a Catch-22 situation as they try to resolve the problem.

Family Planning

For how long, at the present rate of population growth and pace of improvements in crop yields, population growth *per se* will *not* be an issue for humankind to see that all are fed adequately only a fool would hazard a guess at. Population growth must be slowed. Chandler (1992) offered no suggestions as to how population growth would be curbed. He contented himself with making the dire warning that zero growth will become necessary for human survival!

The resources of the world cannot support uncontrolled, rampant population growth. Steps must be taken that are certain to be unpleasant and unacceptable to many.

- Government policies that provide subsidies to encourage large families must be discarded and replaced with other social programs such as job creation programs.
- Religious doctrines that discourage or prevent the practice of birth control must be overturned, rescinded, or relaxed in order to allow the teaching and practice of birth control measures.
- Family planning and the tools to support it must be available for every family that wants them.

The impediments to family planning must be overcome in order to slow population growth.

Land Reform

Land reform is needed. Arable lands should not be paved over for urban sprawl to prevail. Urban or industrial development of arable lands must be curbed. Nor should forested lands be destroyed and the environment threatened for the production of economic cash crops at the expense of food crops and beef production. Cash crops grown by developing nations used for the benefit of developed nations must be balanced against the need for the production of indigenous food crops to feed the local population.

Land reform is not an issue to be resolved by a technocratic government dominated by science. It can only be resolved by the human experience, by emotional, even irrational, and humane principles.

Respect for Indigenous Agricultural and Food Cultural Practices

It could be argued that the last point, demonstrating respect for the agricultural practices and food cultural habits of other peoples, is one that technology might have a role in resolving. The agriculture and food practices of one dominating society or culture (i.e., usually seen as being the developed world) must not be allowed to prevail over those practices of all others (the developing world). That is, an agriculture of one society must be valued within its cultural, traditional, and geographic environment as much as those agricultural practices developed elsewhere. Traditional foods prepared from indigenous crops and animals must be respected. Many see a danger that the use of genetically modified crops will lead

content, for more of a particular phytochemical, or for whatever characteristic, should not be meant for use in underdeveloped regions.

Such practices do not encourage production of local indigenous crops. Local crops and the products made from them have, already, a history of use by local peoples. They are productive in the local terrain and local farmers know how to grow them. In addition, these local crops and animals and products made from them often are more nutritious than crops, animals, and products introduced from other cultures.

Such practices of introducing foreign agricultural crops and animal husbandry practices may mean a loss of a diversity of germ plasm local to the regions into which they were introduced. Should not research be directed to modifying the characteristics of the indigenous crops which the local population knows and uses in cooking?

To continue this point but on a social and cultural note, food habits and practices of one culture should not be allowed to intrude and dominate those of other cultures. Dog and cat meat are delicacies in Korea and Thailand, and half the cost of pork and chicken. Thailand has recently bowed to pressure from Western animal rights groups to ban these delicacies. The Western groups threatened to boycott Thai products. This is arrant and arrogant colonialism.

One can imagine the hullabaloo that would arise if, for example, India refused to permit the food products of countries which slaughtered cattle for food or to do business with such nations that did. Or if nations who revered the horse refused to allow French wines into their countries. And finally, the North American public was treated recently (July/August, 2000) to the ridiculous report in the newspapers that a judge in New York had ruled that regulations for kosher foods were unconstitutional and represented an intrusion of religion into government!

Azam-Ali (2000) reports on a workshop on the importance and role of traditional food products (October 20, 1999, under the auspices of the Appropriate Food Technology Subject Interest Group of the IFST, U.K.). The values of traditional food products were listed by M. Battcock as follows:

- They provide sustainable livelihoods for many in small food manufacturing facilities.
- They represent appropriate and sustainable products suited to their regions and use natural materials and modest technologies.

■ These foods often have a role in religious, social, and family celebrations.

Food practices of other cultures must be respected.

In summary, all the above are issues that greatly influence the world's ability to feed itself and future generations. At present, these problems are left to agricultural economists, religious leaders, and governments to muddle through, but they cannot for long be left to their handling. They will not be resolved by technology. Nor will they be solved by the politics of government. Pricing policies are political issues beset with vested interest groups which provide political pressures and constraints. Though they were not intended to, many agricultural pricing policies have discriminated, in fact, against agricultural producers. Religious taboos against birth control must be relaxed. Governments must find equitable ways to administer agricultural policies and priorities, raise standards of living, and encourage and support family planning and humane birth control measures.

These are tasks, endorsed by many within the food microcosm. They are far beyond the technological capabilities of the food microcosm to resolve. Nevertheless, should there not be a united voice of food technologists (not technocrats) speaking out for the encouragement of diverse agricultural and culinary practices and of research into these crops from other agricultural systems to improve them for local or regional consumption? It is to be hoped that Ausubel's second thesis (1996) with technology's intrusion, i.e., that mankind often does not use technology wisely, does not come true.

Unfortunately, the task here is to concentrate on those problems that food interests can contribute to the solution.

"Food, Glorious Food" (from the musical Oliver)

Humans are omnivorous. What humans eat is derivative of the environment; that is, it is the result of climate, geography, soil type, population densities, arable land space as well as social, cultural, historical, political, and economic influences. People's tastes in foods and the various taboos they harbor respecting foods depend largely on what they grew up eating within their family and cultural traditions. The various food traditions see people prizing one food as a delicacy; for example, calves testicles lightly

favored in Southeast Asia; or geoduck, a large clam prized in Asia, or roasted ants enjoyed as a snack in Colombia. Food habits are not easily or quickly changed — nor should they be.

More Food through Increased Crop Yields

The Green Revolution brought enormous increases in yields of agricultural crops. India and China both became self-sufficient in wheat; India, in fact, became an exporter of wheat. All this was done largely, but not entirely, by traditional methods of cross-breeding to produce better varieties of crops which produced higher yields with often higher protein or starch content. These increased yields of crops also required the extensive use of chemical fertilizers, pesticides, and herbicides. The latter are not always available to poor farmers or, if they are, they are costly for them.

It is here that opposition begins to grow. Those who advocated green farming (or organic farming) protested. So also did the environmentalists. Neither liked the heavy dependence on chemicals, pesticides, and herbicides. Others, environmentalists among them, worried that the new varieties of crops, especially those created by genetic modification, would put the world in danger of being forced into an agricultural system dominated by monoculture. That is, the world would become dependent on fewer and fewer varieties of crops and those crops would be those demanded by the developed world. The result that is foreseen by some agronomists is a loss of the germ plasm of indigenous species. An ancillary worry was that the seeds for these varieties would be under the tight control of a few multinational chemical/biotechnology companies.

How much further can crop yields be pushed? Some scientists say the biological limit has already been reached (reported in MacKenzie, 2000). If a biological limit has been reached, then more land must be diverted to high yielding cereal crops. More land to agriculture means the use of more chemical fertilizer, supplemented with composted materials in developing countries, and the use of more pesticides and herbicides.

Future world demands for food cannot be met by organic farming techniques using green manuring plus compostable material and without the use of pesticides. First, it is a much more difficult agronomic practice to properly administer. Second, and more importantly, there is simply not enough compostable material available to sustain such a program. Products from organic farming can only be produced to cater to those few customers

To make available the yield potential of the newly developed varieties of engineered crops (genetically or conventionally engineered), agronomists will have to carefully select those combinations of (largely chemical) fertilizer mixes, water, and pest and weed control best suited for these crops. This will have to be done, and optimized, for each agricultural niche according to its soil, intensity of light, and temperature conditions in that niche.

It is clearly not an easy task and is one beyond the resources and feasibility of most farmers with small landholdings in developing countries. They combine smallness of scale and poverty; they cannot afford the technology. According to Arulpragasam (1985) 80% of the farmers in the developing world can be classified as subsistence farmers.

A major technological problem will be in developing *indigenous* (to the regions of developing nations) high yielding plant varieties suitable for arid, semi-arid, and even rain-fed lands. Arulpragasam (1985) suggests that research might be better directed to the development of varieties suitable to inferior lands which represent 80% of the land than directed to more productive varieties suitable for the 15% of the land which needs irrigation.

Technology should emphasize improving indigenous crops which are in danger of being neglected and which survive well under local conditions without the need for intensive, high technology agricultural practices.

More Food through Intensive Animal Husbandry

Factory farming of animals has been practiced for many years. Livestock, including pigs, poultry, calves, and even fish have been bred for better feed ratios, i.e., to use feed to flesh more efficiently. They have been raised in confined, carefully controlled environmental conditions. The environmental concerns and ethical and moral issues raised by animal rights groups about intensive animal farming have been discussed in earlier chapters.

These issues are still unresolved. The environmental concerns are amenable to correction — the technology is available — but putting the technology into practice will be expensive. Costs for factory farmers will rise if the technologies are put into practice and ultimately food costs must rise. It will require government intervention to enforce adherence to the pollution standards that are set. Both rising costs and law enforce-

through biotechnology and by the administration of medication. These techniques have involved (MacKenzie, 2000):

- Breeding larger animals (cattle rather than pigs and chickens) for feed efficiency rather than weight gain. This would reduce the demand for grain.
- Designing feeds for more efficient use by animals. As a result they produce less manure and hence are less polluting. Crops such as maize and forage plants, for example, are being developed which contain less phytic acid resulting in less phosphorus supplementation and better protein digestion.
- Maize containing greater amounts of lysine and methionine has been developed.
- Injection of DNA encoding for controlled release of a growth hormone allows pigs, for example, to growth faster on less feed.
- The use of adrenalin-like drugs can make animals lay down muscle instead of fat, which also results in better feed efficiencies.

The above are not, however, technologies suited to the small- or even medium-sized landholdings which are the majority of the landholdings in developing countries. For these farms, technology should be based on existing farm conditions, existing knowledge and resources, and on low risk technologies.

The uncritical application of technologies from what Arulpragasam (1985) has called the "demand side" characterized by abundance of land, capital, scarcity of labor and technology can have unfortunate, if not disastrous, results for small farm holdings.

Nor are the above technologies acceptable to animal rights groups. To avoid confrontations and conflicts with animal rights groups in the future, all the protagonists including primary producers, food manufacturers, retailers, and consumers must come to an understanding with all the activist groups. As Fennema (2000) points out there will be confrontations as opposing sides fight for their beliefs. Somehow the gap between extremists and middle-grounders must be bridged (Anon., 1998a). It will not be bridged by scientists and technologists declaiming that the public must be educated in science. It just might be the technocrats who need to be educated and sensitized to the needs of the public.

of food available for humankind although many of these foods may not be acceptable to all cultures. Vietmeyer (1984a) and Arulpragasam (1985) both suggest that small animal husbandry is ideal for the landless and for those with small landholdings.

Vietmeyer suggests animals that can be raised for food by the landless include

- Chickens which stay close to the house (and forage for themselves), pigeons (also a forager) and quail.
- Guinea pigs which have served as food for centuries and were a delicacy of the Romans and presently are used as food in Peru.
- Bushmeats (rodents such as the grasscutter and Nigerian giant rat) and the Blue Duiker (small antelope). They are the subjects of domestication programs in Africa. These animals, indigenous to their regions and to their environments, can serve as high quality, nutritious meats.
- Reptiles are also excellent sources of meat. Iguana farming is already being investigated.

Micro-livestock is another source of quality protein. Snails have been raised for centuries. Their care and feeding are relatively simple and require little land resources. Entomophagy also offers opportunities as a supply of good quality protein. It is only in North America and Europe where the squeamishness of eating insects prevails. In much of the rest of the world, insects are not only accepted as food but are considered to be delicacies.

Herein is a difficulty. Such foods as these should not be looked upon with disapproval and disgust by dominant food cultures which do not esteem these sources of food. They are accepted, often prized, foods in some societies. More funding should be made available to explore such sources and to investigate small animal husbandry for development in those cultures where they are prized.

There are under-exploited food crops which serve as sources of food (Vietmeyer, 1984b). Many of them are being investigated for crop improvement and export. Grains such as quinoa and amaranth are available; millet and sorghum are enjoying a resurgence. Fruits such as naranjilla, passion fruit, and tree tomato can be purchased in many green grocers. Vietmeyer lists several root crops known to the Incas: oca, arracacha, maca, yacon,

The desirable feature of such crops, and there are many others around the world, is that

- They are nutritious.
- Many can grow in climate conditions or soil conditions that are difficult for many of the better known crops of developed regions which scientists have tried to introduce.
- Because they are indigenous to their regions there are peoples who know them, recognize their properties, and know how to prepare them.

The danger is, of course, that these crops might be lost either by dominance of transplanted crops and agricultural practices popular in the developed countries or by disregard, or even disparagement, by consumers in developed countries or by lack of commercial interest and development.

Blain (1987) quotes a statement made some 20 years before her article was published:

> *"There are some 50 000 vegetable food species available worldwide, but mankind survives on merely 50 to 200 of them: this is not only insufficient, it is also dangerous."*

She describes neglected crops such as quinoa and cassava, and their many applications in food products. In addition, amaranth, teff, fonio, yams, sweet potatoes, and plantains are suggested as possibilities for development. Her thesis is that encouraging and developing these local food crops are more beneficial than subsidizing costly food imports from developed countries.

Fighting the Current: Forces in the Marketplace

Three important inter-related concerns for those involved in the food microcosm in the third millennium emerged from earlier discussions about relationships among the customers, consumers, sellers, and manufacturers that are found in the varied marketing arenas. These were

1. Dominance and control. Who dominates, i.e., rules, in the various marketing arenas? Who truly controls these marketing arenas? And

on brand loyalty? Whom does it serve, customer or retailer? Will bricks-and-mortar outlets, especially the smaller ones, survive?
3. Security and privacy. Have the tools of market and consumer research become too good and too invasive? How secure are transactions by electronic commerce?

The various marketplaces for food selling and all the players in them are being buffeted by many influences. Competition is intensifying for manufacturers to get and keep customers/consumers. Customers/consumers have more in-store and/or in-home tools at their disposal to be more informed food shoppers and, therefore, more discriminating in their food purchases. Mass marketing has suffered a heavy blow from the power of e-commerce for niche marketing. The tools for consumer and market research are becoming more powerful for marketing and more dangerous to privacy.

"Power tends to corrupt and absolute power corrupts absolutely."

John E. E. Dalberg-Acton, Lord Acton Letter to Bishop Creighton, 1887

Who's Top Dog?

A shift in power occurred during the last millennium (Figures 2.1 and 2.2). The power base depicted in Figure 2.2 is, nonetheless, unstable for reasons which will become apparent shortly. The relationships (Figure 2.2) are changing drastically. The emerging power structure is that depicted in Figure 9.1.

The Primary Producer

The primary producer went from being top dog at the beginning of the past millennium (Figure 2.1) to lowest in the pack (Figure 2.2) at the beginning of the third. At the beginning of the past millennium, the primary producer was in many instances the manufacturer and the retailer. There was a gradual specialization of these functional roles of the primary producer during the latter half of the second millennium; the roles of manufacturer and the retailer/seller became more distinct. The primary producer has not shifted at all in power or importance

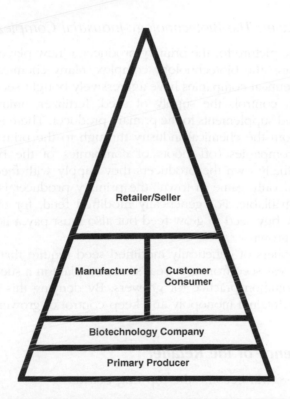

Figure 9.1   The New Relationship among the Protagonists of the Food Microcosm

products. Producers, at the bottom, are stuck with getting rid of genetically modified crops presently on their hands.

The Manufacturer's Power Base

Manufacturers now contract with primary producers dictating to them:

- What to produce, the quality that is desired, and the price that will be paid
- What agronomic practices (timing of pesticide and fertilizer applications) will be used in producing the raw material

The New Player: The Biotechnology Industrial Complex

To worsen the picture for the primary producer, a new player has entered into the picture: the biotechnology complex. Many chemical-biotechnology-pharmaceutical companies have aggressively bought seed companies. This complex controls the supply of seed, fertilizer, animal feed, and medicated feed supplements to the primary producer. There is real, vertical integration from the chemical industry through to the primary producer. Many feed companies (off-shoots of companies of the biotechnology complex) virtually own the producers they supply with feed. Since they represent the "only game in town" the primary producer is restricted in the options available. For genetically modified feed, for example, they must not only buy seed to grow feed but also must pay a license fee per acre of crop grown.

Some suppliers of genetically modified seed require that growers not carry over excess seed from one year's crop to plant in a succeeding year. This was a common practice for growers. By denying this activity, seed companies maintain a monopoly and keep control of growers and prices.

The Emergence of the Retailer

Retailers/sellers will become the top dog because

- They control the marketing arenas. They control their structure, their location, their form, and their services, including the new food e-tailing.
- They control what products are displayed, how and where they are displayed and promoted; and
- They control promotional pricing.

The retailer also controls the manufacturer, perhaps in only subtle ways, but the control is there. For example, this is how one manufacturer described just-in-time-delivery practices: "You will send me what I want on my store shelves in the quantities I want when I want it and at the price I want." Indeed, many retailers are linked electronically to their manufacturers so that as product is purchased, replacement orders are transmitted automatically to the manufacturer. For manufacturers this means that there will be more frequent small deliveries and, consequently,

customers. By analyzing traffic flows and how displays or promotions attract customers they can exercise some control over customer purchases. They are also in a position to survey individual customer groups (seniors, males, females, young, with families in tow) as they shop: Who do they buy for? When do they buy? What do they buy? and, to observe and analyze the products and combination of products that are purchased. In this sense, retailers have available the power that customer research can give them. They can determine the most profitable items and combinations of items to stock and place them for maximum exposure. They can prompt impulse purchases of related items.

Retailers, then, can wield considerable power by

- Making demands of food manufacturers concerning price, delivery, packaging, advertising, and promotion of products that they display.
- Controlling or greatly influencing what products are displayed and how they are displayed. They control the introduction of new products with their demands for slotting fees.
- Manipulating or being a major influence through selling techniques on what customers can/will purchase.

Can this power of the retailer be abused or will it be abused? Some consumer groups believe both questions can be answered, "Yes." Whether this abuse is intentional or done under the guise of serving the customer better is a moot point.

Such customer/consumer information that the retailer can command may maximize the earning power of store space and it may satisfy or be a convenience for the majority of customers for the majority of their purchases of the most popular items. However, this cannot satisfy all customers for all their purchases, i.e., it cannot appeal to all tastes. An anecdote illustrates this point:

> I followed a customer down the stalls at a farmers' market in corn season. She became increasingly vociferous as she progressed past the stalls. Finally she was speaking out loudly to the vendors: "Hasn't anyone got corn that tastes like corn? Do you only grow 'peaches and cream' corn" [a newer variety of corn that is earlier maturing, sweeter, milder flavored and softer textured]. Nobody did. One vendor told her it wasn't popular and no longer available. (It was available.)

the absence of particular brands of packaged biscuits and cookies, and flavored, powdered milk-based drinks. They were all still available; they just weren't popular in that region so were no longer stocked by the local stores. Retailers will not stock slow sellers or at best will not give prominence to slow selling specialty items. In short, the retailer now is able, in effect, to orchestrate the business of selling such that the customer will (must) purchase what the retailer wants the customer to want to purchase according to the retailer's customer statistics.

Is the application of such information becoming too manipulative? Customer purchases can be analyzed for an infinite amount of information and the techniques for doing so are becoming more sophisticated. Is all this a convenience for the customer or for smart retailing? The retailer is, after all, in the business of making a profit and satisfying shareholders.

With improving technology, retailers are finding the customer ever more predictable and tractable and this may in the long run provide retailers with the power of manipulation.

E-commerce and Food E-Tailing

All national food retailers have moved into this arena to some degree with Tesco™ stores in the U.K. claiming to be the largest e-tailer in the world. The potential for marketing their own products has even allowed some manufacturers to retail their own products.

Are there some bubbles to be burst regarding e-commerce? The most obvious burst bubbles have occurred already: Most virtual food and non-food outlets have yet to show a profit and steady growth. There are almost daily reports in the business news section of newspapers announcing that one of these outlets is looking for an infusion of new money, is closing, or is going bankrupt.

What are some of the questions and concerns for customers, manufacturers, and retailers? The following are some that have been reported:

■ How will governments react in this as yet unregulated arena? They may step in with closer regulation of advertising claims and promotional activities. After all, if Main Street stores have to pay property taxes should not virtual marketplaces also have some tax levy? How will sales taxes be collected?

■ How will food manufacturers with their carefully nurtured brands

location but several locations nearest the customer displayed? The customer can fill a food order with the best product at the lowest price from a multitude of stores. Company and brand names lose their advertising impact; it will be lowest price for best quality. Brand loyalty might suffer.

- What will be the impact of food e-tail on customer service? On-time delivery is a problem where perishable food items are involved. The customer must wait for delivery (be home at some designated time) or arrange to pick up food at work – a service some companies are providing – or at the nearest store. There have been problems when spoiled or damaged foods or recyclable bottles and cans which were purchased at (i.e., through) a virtual store have been returned to a Main Street store.
- Customers using food e-tailing can purchase food products wherever such electronic service is available. Any customer with access to a computer and the Internet can view the catalogues of native or regional gourmet foods from all over the world and make purchases online. Sales of such regional French delicacies as duck and goose foie gras, truffles, and wines can be ordered from the Perigord region in France. Other Web sites have been developed for the sale of other ethnic gourmet delicacies to provide the customer with a broader range of products than could be obtained before e-commerce.
- Will auction selling become a feature of food e-tailing? It is already a well-known feature on the Internet. It has become a fact with the sales of airline tickets. Auction sales are well-known for the commercial sales of cut flowers, for example. Anyone who has visited a farmers' market knows full well the subtle form of auction selling that goes on. Produce prices drop steadily during the day and are at rock-bottom levels for bargain seekers when the market closes at night and farmers are packing up to leave. Would a form of this be applicable to food e-tailing?
- Will the result of electronic selling of food increase or decrease competition? Any increase in competition would produce benefits for customers with reduced prices and increased services.

This novel selling arena instead of being a Wonderland of opportunity, as many business columnists and journalists have described it, has the

newspaper spread that I saw — and it must be described as that, promotional — pictured a young married professional woman with two children under 5 years of age. She hated shopping especially with two children in tow. The family was a two-income family; the father was also a professional. They had a daily babysitter who was available to receive and put away the groceries when they arrived. Certainly, food e-tailing was a convenience to them.

The other target for depicting the convenience of electronic food shopping is the incapacitated or wheel-chair-confined individual for whom this access to food shopping would be helpful.

It is expensive and can only be expected to add to the cost of food. Most stores charge $8 to $10 to deliver an order. Financial analysts have estimated, however, that the true cost of a delivery is closer to $20. How long retailers will be able to underwrite this cost without either charging more for deliveries or absorbing the cost by raising food prices throughout their entire product line is anybody's guess.

How e-commerce will conflict with mass marketing has been mentioned previously. E-commerce opens up the possibility of directly influencing a unique customer, one who is no longer faceless and nameless.

Security and Privacy

E-commerce is still, in the minds of many, not secure. Customers are reluctant to provide financial information over the Internet. They are worried about the security of such information and their concerns are heightened when news of hackers who break into Web sites and publish credit card numbers is bruited about. This aspect of electronic security is beyond the scope of this book.

Security of financial information is not the only concern. Other concerns besides the privacy of personal and financial information given through e-commerce transactions are

- Profiling of customer purchasing habits (products purchased, for example) collated through credit and debit card information in both e-commerce and in-store transactions.
- Sales of club and association membership lists, magazine and journal subscription lists, etc. and the sales of such data to market research companies.

"I basically equate cookies to the notion of a store being able to tattoo a barcode on your forehead, and then laser-scan you every time you come through the doors."

S. L. Garfinkel cited in http://www.anu.edu.au/
people/Roger.Clarke/II/Cookies.html

Techniques to gather customer/consumer profiles are becoming more sophisticated. Unfortunately, technology to combat them is not. Market researchers want them and use the subtle argument that their use serves customers better and saves customers having to resubmit personal information each time they register online. Governments have shown an interest in these profiling techniques for "tracking" its citizens.

Customers/consumers may find they have to live with cookies and other profiling technologies, or adopt blocking techniques against their use to protect themselves, or lobby for legislation against the distribution of any personal data without their consent.

Education: Whose Responsibility?

Education was a recurring theme through many of the preceding chapters and in earlier sections of this chapter. It arose during reviews on food safety throughout all stages of production, food processing, handling, distribution and storage, retailing and in-home preparation and handling; during discussions on health and nutrition; during an examination of the concepts of safety and risk assessment; and in explorations of legislative dilemmas.

Education of customers and consumers as well as education of factory floor personnel, distributors, and retailers is required to reduce the risks of unsafe food in marketplaces and in homes, to enhance the quality of life of world populations, and to feed future generations. Training is essential for all within the food chain right down to the stock clerk filling shelves in a grocery store. The problem is, of course, such education for safety will cost money.

The onus for education does not rest solely within the food chain. Educating consumers must rest in large part with some level of government.

How necessary is education concerning food, its nutrition, its preparation, its potentially fragile nature? The following episode took place not

> *A young woman who had grown up with my own children asked me a question. She had baked a carrot cake which had fallen "as flat as a pancake." We chatted about some possible reasons for this failure. During our conversation we were joined by another young lady. She firmly declared that the cake failure was probably because my young friend was menstruating at the time!!! It was more disturbing to learn later that the young lady with this 'biological' explanation was an interning medical student!*

That such folklore about food should be prevalent in 1999 when this incident occurred is appalling. That it should exist in someone who will be looked to as a health specialist, as someone who will advise others on health matters that very likely will involve food and nutrition is terrifying. Can anyone then wonder why there is such controversy and misinformation about food and the food industry in general?

Educating the Public

The public is the customers and consumers. They range from the very young to the elderly. They may suffer from a variety of ailments and handicaps. They know less and less about food *per se* and the workings of the food microcosm in general. Who will educate the public?

It will certainly not be the scientists: food, nutritional or otherwise. The public does not have full confidence in science and scientists when it comes to issues of safety and health concerning its food supply. The public resents its exclusion from the decision-making process and resents the disdain shown by many scientists for its, the public's, concerns. The public is confused by the often divergent views expressed by experts on safety, health, and diet topics.

Governments certainly have the resources to educate; they supply funding for education. Primary and secondary schools in the developed world, however, are hard pressed now to provide the requisite education to their pupils plus ancillary services to pupils with learning disorders, behavioral problems, and physical handicaps. They do not have the funds from their governments to teach food-related subjects nor do teachers have the skills to teach food, health, and nutrition.

Primary and secondary school curricula need government funding to provide basic education in nutritional sciences, in the properties of foods, safe food preparation techniques in the home, personal hygiene, and

This leaves industry as the educator if government will not or cannot support education. Two questions are raised: Can industry be entrusted to do the job? and Does industry want to do the job of educating the public? About the latter question, a marketing professional once, several years ago, told me he would rather be a Daniel thrown into the lions' den than be given the task of educating the public. Opinions expressed by others have confirmed this aversion to the task of being an educator.

The marketing side (development of strategies for communicating with the public) of the food manufacturer is not suited for education of the public, nor should the food industry be expected to perform this task. They may provide complete nutritional information about their products, ingredient statements, recipe booklets, and meal planning suggestions. Beyond that marketing's role in education is very limited and not to be looked upon as an educator but only as that of a communicator of the desirability of a product.

Education and Training in the Food Service, Manufacturing, and Retailing Industries

Reasons for the food industry's conservatism in its adoption of advanced technology were discussed in Chapter 6.

Demetrakakes (1998) cites reasons why the food industry does not have the *skills* to move quickly in adopting more sophisticated technologies:

- The food industry is often a last employment choice industry. Consequently, the industry generally gets workers with low basic skills. This work is all that is available to such workers.
- The working environment of the plant floor is frequently hot, humid, dirty, and noisy or cold, dirty, and noisy. Skilled workers prefer not to work in such an environment.
- Food plants are often in agricultural areas. They draw upon agricultural laborers who are frequently seasonal migrant workers with little formal education and may not have basic language skills. Many workers may not speak or read the working language in the factory.
- Agricultural work is seasonal; manufacturing agricultural products is seasonal. It is difficult to keep skilled workers if they cannot be given year-round employment. Off-season work is largely confined to maintenance. Such work is rarely challenging enough to keep

The above are generalities. They do not apply to all plants in all geographic locations. Food plants in large urban centers are more successful in the caliber of employee they can hire because these areas offer greater social, cultural, educational, and employment opportunities for workers and their families. They must also compete with the better wages and working conditions that other industries offer.

The dilemma, then, is that the food industry is left with few alternatives (Demetrakakes, 1998). For example, it can:

■ Eliminate the need for the involvement of workers and their labor, i.e., automate as much of the process as possible. There are serious consequences to this.
■ Introduce systems that can overcome the workers' deficiencies and/or enhance the skills of workers through training. Such systems must make the workers more productive resources for their companies.
■ Select employees by carefully screening them prior to hiring.

This third alternative is quite acceptable for selecting candidates if there is a large, skilled labor pool available. This is not the situation in small agriculturally based farming communities.

Elimination of labor through automation of certain operations and processes puts food manufacturers into a 'damned-if-they-do-and-damned-if-they-don't' situation. Any sophisticated equipment employing advanced technology requires more skilled staff to operate and maintain it. Workers lost through automation collectively were a wealth of experience that is now unavailable to the company. These people knew how the system was meant to operate if production failed: If automated production operations broke down or if glitches developed in computer-controlled systems they could operate the plant manually (Demetrakakes, 1998). They are now lost to the company.

The food industry, including the service and retailing sections, must adopt training techniques for its employees. The benefits could be enormous for the safety and quality of foods and for employee health and stability:

■ Personnel with upgraded skills would be more cognizant of concerns for hazardous processing and handling steps that influence

- Personnel who were more skilled would be able to perform their jobs more efficiently, with less down-time, better product quality, and improved production performance.
- The company would have a core of trained and trusted employees from which to draw workers for more responsible managerial positions.
- Employees would have a boost in morale, and see the work as career-oriented.

Several elements can be factored into a need for reducing concerns about food safety:

- Workers must be made aware of and recognize food hazards of public health significance or conditions leading to situations creating possible health hazards whether these be in the factory, the canteen, restaurant kitchen, or in the home.
- All workers handling or preparing food must know and use all possible means to avoid these hazards or minimize their occurrence.
- They must recognize the economics of safety. Loss of food quality through mishandling and poor preparation or illnesses or deaths resulting from eating tainted food are major economic losses that can result in lawsuits for individuals and companies, factory closings, and loss of income.

To educate the public about proper food purchases, good health and nutrition, and about the safe handling, storage, and preparation of food requires money. For food manufacturers and food retailers to prepare and present safe food requires training of staff in the principles of GMPs and HACCP programs at the factory and at retail levels. This, too, has a cost factor. Some responsible body or bodies must assume responsibility for the education of the general public respecting nutrition and food hygiene and educate food preparers in safe food handling.

The New Nutrition: Blessing or Curse?

The challenge in the third millennium will be in distinguishing between blessings and curses that the revolution in nutrition will bring about. On the one hand there is the promise of health benefits from phytochemicals

If the benefits do materialize as safe preventive approaches to maintaining one's well-being, these attributes must be communicated responsibly to consumers. The natural tendency on the part of both the consumer and manufacturers is to fortify or over fortify products with the beneficial phytochemicals — it is the 'if one pill is good for you two will be even better' school of thought. Such thinking might prompt legislative intervention respecting labeling, levels of permitted addition to products, or restricted applications in foods.

Personalization of diets might very well become a reality. It presents some interesting, indeed, even far-fetched, possibilities. Consumers, who through the knowledge of their own family histories or even based on their personal genetic mappings, believe they may be susceptible to certain genetically linked disorders, have the opportunity to design preventive diets. That is, they would know which foods they should maximize in their diets, and which to avoid or minimize in their diets to reduce the likelihood of their genetically linked disorder from being expressed.

There is another side to this. Might some foods have to be labeled with a warning that if one is of a particular genotype one should avoid these foods? Or the food be allowed to bear the statement recommended for those with such-and-such a genotype?

Safety and health benefits need to be clearly defined before the potential of functional foods in the diet and as adjuncts to food products can be fully realized. It would be distressing if the hopes bruited about for these foods and the phytochemicals in fruits and vegetables proved groundless. The public's already shaky faith in nutritional science would be shattered.

Food: On Being an Environmentally Responsible Corporate Citizen

The food microcosm is in an impossible catch-22 situation. Producing food in the quantities and at the quality customers worldwide want causes pollution. Feeding the world's growing population will cause pollution. This has been discussed. The problem is to find and apply solutions. Some were described; all are expensive.

The problem for the new millennium will be to clearly delineate which of the waste reduction procedures are truly most cost effective and are themselves least damaging to the environment. Solutions are apt to run

Genetic modification of plants offers many promises of answering some of the problems respecting world food and pollution. There were, however, both strategic and tactical errors made by the biotechnology complex in their introduction of these crops. The modifications were not shown as customer/consumer benefits but only demonstrated as conveying benefits to the producers and to the biotechnology companies themselves. Environmental concerns, which ought to have been anticipated by the biotechnology firms with the full vigor that has been demonstrated, were also not addressed openly. Some questions must be posed here:

- Why were the results of environmental trials not made public and debated in a public forum for all concerned to review or contribute to?
- Why were the purported benefits of these crops to the consuming public or to poor farmers in developing nations not discussed?
- Why the unseemly (to the public) haste to market these crops before long-term human feeding trials had been done? What toxicity testing was performed on genetically modified crops?
- Why the hasty retreat from development of crops with the "killer" gene?

The promise of genetically modified crops was dropped suddenly on an unsuspecting public, one made up of customers/consumers, environmentalists, scientists who were pro- and anti-genetically modified foods, and interest groups ranging from pure food addicts to back-to-nature advocates. There was no preparation of these groups through the media. The battle for the acceptance of genetically modified foods will not be won in scientific journals but in the various media available to the public. Scientific papers written by scientists supported by the biotechnology complex will always be suspect. In the public media, open discussion by concerned people, by scientific journalists and commentators, and by others in the food microcosm might go a long way to paving the way for better acceptance.

Safer Food

Food cannot be made *absolutely* safe. Legislation and its support systems of standards, sampling and analysis, and inspection cannot make food

272 ■ Food, Consumers, and the Food Industry

How then can the integrity of the food chain respecting both safety and quality be assured? It cannot. There are no fail-safe mechanisms or systems that can supply this assurance. The most that can be done is to reduce the opportunity for the occurrence of known hazards to 'acceptable' levels (i.e., levels known to not cause incidents of public health significance). Assistance in maintaining the integrity of food products can be improved by education of:

- The public in the safe and proper handling and preparation of foods, in the fragility of food preservative systems to abuse, and in personal hygiene.
- Food production workers in the adoption of agricultural practices that enhance the safety and quality of food or do not contribute to the contamination of food.
- Workers in the food manufacturing industries in the safe and proper post-harvest handling, processing, and post-preservation handling of food and in personal hygiene.
- Workers (including shelf stockers, meat cutters and butchers, and produce handlers) at the retail level in proper, safe storage, handling, and display of sensitive food products.

Many people are allergic to foods or have unpleasant reactions to the consumption of some foods. The varieties of foods causing these undesirable effects are legion and the nature of the unpleasant reactions highly variable. Efforts to remove the allergens found in foods would be better directed toward the development of rapid detection methods to find the presence of major allergens and so allow consumers to test foods of questionable provenance or preparation on site with immediate results. Removing allergens would protect only those with allergic reactions and do nothing for those with non-allergic food reactions. It would also increase the likelihood of error in the mixing of non-allergenic foods with allergenic ones during food processing.

The responsibility must remain with those with allergies or sensitivities toward particular foods to avoid these foods. To do so, however, these consumers must have the tools to identify such foods. Labeling is certainly not sufficient: "May or may not contain (pea)nuts" can hardly be considered an adequate warning on labels. Not all eating situations provide ingredient descriptions. Necessary to safeguard such sensitive people in

New technologies for the removal of or desensitizing of the triggering components in the processing of foods and thereby providing protection of exposed consumers.

Rehabilitating Science

There will always be controversy in science. As a consequence, science journalists will be quick to exploit such differences of opinion because these *are* interesting news items. Vested interest groups will always be eager to select those informed opinions supportive of their causes. The sciences related to food, its production, its processing, and the associated nutritional sciences seem particularly prone to controversy.

Governments realized that strong research programs plus a well-educated work force stimulated industry and promoted prosperity. They supported studies on their nation's resources in agriculture, minerals, and other natural resources. The corporate world is now taking over from governments as the chief sponsor of research as governments cut back their financial support. Indeed, universities are now in danger of becoming the corporate research arm of industry. Universities no longer seek pure research but must direct their efforts to applied research that will attract corporate funding.

All funding has ties. Each change from church to government to corporate sponsorship of university studies has involved ties of one sort or another. One simply did not offend a patron, no matter who that patron was. All sponsorship of research, especially corporate sponsorship, must be declared clearly and unambiguously. Sponsorship does introduce a bias, however unintended.

It has been suggested by the academic community that peer review of publications would eliminate 'bad' science. This is certainly a step in the right direction. Critics of peer review, of whom I've heard a number, do consider this, too, is flawed. Their criticism is that the peers, the keepers of the current dogma, are asked to review that which may be counter to the accepted dogma. Peer review might be a deterrent to ideas that do not conform to accepted theories. Peer review did play a restraining role in the acceptance of the prion as a factor in disease and the *Helicobacter pylori* as a factor in stomach ulcers. Current thought had centered on stress and spicy foods. A similar fate of slow acceptance of ideas counter to current thinking has befallen the emergence of nanobacteria and the

quality of education in the pursuit of studies that may be contrary to the established dogmas in pure sciences, in mathematics, the liberal arts, and philosophy.

Scientists suggest that if the public were better trained in science they would be more understanding of the TRUTH; that is, the scientific truth as they, the scientists, convey it. No doubt there is great value in increasing the science content of primary and secondary school curricula. To regain the confidence of the general public, scientists must accept that science can explain the material world but often explains it imperfectly. Perhaps, on the other hand, scientists should be more educated in ethics, philosophy and logic, religion, the liberal arts, and politics that they may understand the real world, humankind's world.

The Challenges: A Summary of Confusing Issues

The salient issues to confound food technology and all within the food microcosm in the third millennium and particularly in the 21st century are many and varied. There will be many solutions offered and many will be disturbed by some of the solutions that are offered. Resolution of the issues is difficult to crystallize when there is such a complex interweaving of influences affecting them. Governments are involved. Their influences on food activities and the influences driving governments were depicted in Table 8.1 and in Figure 8.1, respectively.

Consumers determine what is desirable; customers determine what is purchased. Both are influenced by price, by advertising and promotional materials, by the availability of products where they live, by their own ethnic and cultural backgrounds, by their religion, and by their food budgets. Their demands and their tastes in foods must be met with safe, nutritious foods.

Agriculture is influenced, in its turn, by geography and weather. Geography is a knowable element and no one has yet been able to do anything about the weather. The topography of a country together with climate determines what crops can be grown or what animals raised. Only Herculean efforts by biotechnologists will allow the production of atypical, genetically modified crops and animals able to be raised profitably in regions where land and climate are suboptimal. Such efforts bear the consequences of greater costs which must be borne by customers, man-

needed to feed the world will put severe pressures on land and water resources and inevitably cause pollution. Technology to prevent pollution is necessary.

Manufacturers are challenged by the need to manufacture safe, nutritious, and desirable products in volume and at a price the customer will pay and which their marketing personnel can successfully market. They need an assured source of raw materials at a reasonable price. They must also cope with a changing technology that provides opportunities for new products but also requires, in many instances, new plants and equipment and personnel with new skills and training. Furthermore, they must cope with pollution controls involving water reclamation, solid waste recycling, and odor and noise abatement. The inevitable outcome will be greater costs of production.

Food technology cannot resolve all the issues. Its efforts can only apply to those amenable to a technological resolution. Many problems require the efforts of governments to develop inter- and intra-national policies of cooperative programs for pollution controls, land reform, agricultural pricing policies, and social reforms. There must be greater support for education of the public in food and nutrition sciences at the primary and secondary school levels. The issues of birth control and of the provision of ready access to birth control materials are necessary to curb population growth. These are issues outside the direct control of the elements within the food microcosm.

Progress to resolve the issues confronting food technologists and the food industry "…is, and will continue to be, erratic and fraught with controversy" (Fennema, 2000).

References

Abley, M., Genetically altered food hard to digest, critics say, *The Gazette, Montreal,* A1, Feb. 19, 1999.

Abu-nasr, D., Healthy chips, *The Gazette, Montreal,* W7, Oct. 3, 1998.

Accum, F., *A Treatise on Adulterations of Food and Methods of Detecting Them,* Ab'm Small, Philadelphia, 1820; reprinted by Mallinckrodt Chemical Works, Volume 2, Collection of Food Classics, St. Louis, MO.

Adams, I. M. V., Codex Alimentarius and its work in relation to other international organizations, *Inst. Food. Sci. and Technol. Proceedings,* 16, 81, 1983.

Ahmad, J. I., Free radicals and health: is vitamin E the answer?, *Food Sci. and Technol. Today,* 10, No. 3, 147, 1996.

Ahmad, J. I., Omega – 3 fatty acids – the key to longevity, *Food Sci. and Technol. Today,* 12, No. 3, 139, 1998.

A Lady (*sic*), *A New System of Domestic Cookery; Formed Upon Principles of Economy: and Adapted to the Use of Private Families,* A New Edition, John Murray, Albemarle-Street, London, 1828.

Ames, B. N., Dietary carcinogens and anticarcinogens, *Science,* 221, 1256, 1983.

Andrews, J., *Peppers: The Domesticated Capsicums,* University of Texas Press, Austin, 1990.

Anon., *The Pocumtuc Housewife,* published by The Women's Alliance of the First Church of Deerfield, Deerfield, Massachusetts, Revised Edition 1897.

Anon., *Food of Our Fathers,* Institute of Food Technologists, Chicago, 1976.

Anon., How much good land is left?, *Ceres* 11, No. 4, 13, 1978.

Anon., Worldwide increase in obesity could become disastrous, doctors say, *The Gazette, Montreal,* B15, May 17, 1996.

Anon., A martyr in the making: editorial, *New Scientist,* 160, No. 2164, 3, 1998a.

Anon., Straddling the millennia, *Royal Bank Letter* 79, No. 4, 1998b.

Anon., Millennium in maps: population. *National Geographic* Supplement, No.

Anon., Audits International Home Food Safety Survey, 2nd quarter, 1999a, http://www.audits.com/Report.html.

Appert, N., *The Art of Preserving All Kinds of Animal and Vegetable Substances For Several Years*, London, Black, Parry and Kingsbury 1812; reprinted by Mallinckrodt Chemical Works, Volume 1. Collection of Food Classics, St. Louis, MO.

Arnold, T., Links between second-hand smoke, cancer manipulated says report, *National Post*, A9, May 3, 1999.

Arnold, T., Doctors warned about getting cozy with business, *National Post*, A11, Oct. 20, 1999b.

Arulprgasam, L. C., 'Demand side' technology: an agricultural strategy based on small-farm needs, *Ceres*, 18, Nov./Dec. 27, 1985.

Astorg, P., Food carotenoids and cancer prevention: an overview of current research, *Trends Food Sci. Technol.*, Dec. 8, 406, 1997.

Ausubel, J. H., Can technology spare the earth?, *Amer. Scientist*, 84, 166, 1996.

Azam-Ali, S., Promoting and protecting traditional food products, *Food Sci. and Technol. Today*, 14, No. 1, 44, 2000.

Barnett, A. and Wintour, P., Experiment gone wrong led to mad-cow epidemic: experts, London Observer Service, London in *The Globe and Mail*, A9, Aug. 9, 1999.

Belem, M. A. F., Application of biotechnology in product development of nutraceuticals in Canada, *Trends Food Sci. Technol.*, 10, No. 3, 101, 1999.

Bell, S., Meat-plant inspections get failing grade: audit, *National Post*, A1, Feb. 6, 1999.

Birmingham, C. L., Muller, J. L., Palepu, A., Spinelli, J. J., and Anis, A. H., The cost of obesity in Canada, *Can. Med. Assoc. J.*, 160, No. 4, 483, 1999.

Black, M., *The Medieval Cookbook*, Thames and Hudson, New York, 1992.

Blain, D., Neglected crops, *Ceres*, 20, No. 1, 43, 1987.

Block, E., The chemistry of garlic and onions, *Scientific American*, 252, March, 114, 1985.

Bowle, J., Six centuries of food, *Illustrated London News*, Christmas Issue, 62, 1975.

Brassart, D. and Schiffrin, E. J., The use of probiotics to reinforce mucosal defence mechanisms, *Trends Food Sci. Technol.*, 8, No. 10, 321, 1997.

Braudel, F., *The Structures of Everyday Life: Civilization and Capitalism 15th – 18th Century, Volume 1*, English translation, Harper & Row, New York, 1981.

Braudel, F., The *Wheels of Commerce: Civilization and Capitalism 15th – 18th Century, Volume 2*, English translation, Harper & Row, New York, 1982.

Broadbent, K. P., An "invisible" good, *Ceres*, 11, No. 3, 19, 1978.

Brody, A., Minimally processed foods demand maximum research and education, *Food Technol.*, 52, No. 5, 62, 1998.

Brouns, F. and Kovacs, E., Functional drinks for athletes, *Trends Food Sci. Technol.*, Dec. 8, 414, 1997.

Burk, R. F., Selenium, in *Present Knowledge in Nutrition*, 4th ed., Hegsted, D. M., Chichester, C. O., Darby, W., McNutt, K., Stalvey, R. M., Stotz, E. M., Eds., The Nutrition Foundation, Washington, 1976, Chap. 30.

Burke, D. C., Making British food safe, *Food Sci. and Technol. Today*, 13, No. 1, 12, 1999.

Busetti, M., Fast-forward into functional foods, *Prepared Foods*, 164, 38, 1995.

Caragay, A. B., Cancer-preventive foods and ingredients, *Food Technol.*, 46, 65, 1992.

Chandler, R. F. Jr., The role of the international research centers in increasing the world food supply, *Food Technol.*, 46, 86, 1992.

Chichester, C. O. and Darby, W. J., The historical relationship between food science and nutrition, *Food Technol.*, 29, No. 1, 38, 1975.

Chung, K.-T., Wei, C.-I, and Johnson, M. G., Are tannins a double-edged sword in biology and health?, *Trends Food Sci. Technol.*, 9, No. 4, 168, 1998.

Coghlan, A., Is anything safe to eat?, *New Scientist*, 157, 4, Jan. 3, 1998a.

Coghlan, A., Out of the frying pan, *New Scientist*, 157, 14, Jan. 3, 1998b.

Cohen, L. A., Diet and cancer, *Scientific American*, 257, 42, 1987.

Colapinto, R., Frankenfoods or wonder foods?, *Canadian Living*, 23, No. 6, 50, 1998.

Corcoran, E. and Wallich, P., Overcoming the short-term syndrome, *Scientific American*, 266, 133, 1992.

Corcoran, T., Hard to follow: Is CBC news safe? CBC's week of food scares, *Financial Post*, Oct. 21, C7, 1999.

Cosman, M. P., Fabulous feasts: medieval cookery and ceremony, George Braziller, New York, 1976.

Cowley, G., Cancer and diet, *Newsweek*, Nov. 30, 1998.

Cracknell, M. P., Can producer and consumer interests be reconciled?, *Ceres* 18, No. 5, 15, 1985.

Daniels, R. W., Home food safety, *Food Technol.*, 52, No. 2, 54, 1998.

Danzig, M., Potential pitfalls of industry-university cooperative research centers, *Food Technol.*, 41, 103, Dec., 1987.

Davidson, C., Drinking by numbers, *New Scientist*, 158, 36, April 11, 1998.

Dawson, F., Carcinogenic chemical found in Chinese sauces, *National Post*, A4, Oct. 4, 1999.

Day, M., He who pays the piper..., *New Scientist*, 158, 18, May 9, 1998a.

Day, M., Salt and vitriol, *New Scientist*, 159, 4, Aug. 22, 1998b.

de Graaf, C., Fourth Food Choice Conference, Birmingham, U. K., 1995, *Trends Food Sci. Technol.*, 6, No. 7, 245, 1995.

de la Falaise, M., *Seven Centuries of English Cooking*, Barnes and Noble, New York, 1973.

Demetrakakes, P., Information overload, *Food Processing*, 59, No. 4, 17, 1998.

Derfel, A., No cancer miracle, MD warns, *The Gazette, Montreal*, A1, June 29, 2000.

Dodd, J. L., Incorporating genetics into dietary guidance, *Food Technol.*, 51, No. 3, 80, 1997.

Drewnowski, A., Henderson, S. A., Hann, C. S., Berg, W. A., and Ruffin, M. T., Genetic taste markers and preferences for vegetables and fruit of female breast care patients, *J. Am. Diet. Assoc.*, 100, No. 2, 191, 2000.

Duffy, V. B. and Bartoshuk, L. M., Food acceptance and genetic variation in taste, *J. Am. Diet. Assoc.*, 100, No. 6, 637, 2000.

Duran, G. P., Rohr, F. J., Slonim, A., Güttler, F., and Levy, H. L., Necessity of complete intake of phenylalanine-free amino acid mixture for metabolic control of phenylketonuria, *J. Am. Diet. Assoc.*, 99, No. 12, 1559, 1999.

Durant, W., *The Story of Civilization, Volume 4. The Age of Faith*, Simon & Schuster, New York, 1950.

Durant, W., *The Story of Civilization, Volume 6: The Reformation*, Simon and Schuster, New York, 1957.

Elliott, J. G., Application of antioxidant vitamins in foods and beverages, *Food Technol.*, 53, No. 2, 46, 1999.

Farrer, K. T. H., Frederick Accum (1769-1838) – consultant and food chemist, *Food Sci. and Technol. Today*, 10, No. 4, 217, 1996.

Farrer, K. T. H., To regulate lead, *Food Sci. and Technol. Today*, 12, No. 3, 147, 1998.

Fennema, O., Industrial sustainability: lifting the siege on earth and our descendents, *Food Technol.*, 54, No. 6, 40, 2000.

Forbes, R. J. and Dijksterhuis, E. J., *A History of Science and Technology. Volume 1, Ancient Times to the Seventeenth Century*, Penguin Books, Baltimore, Maryland, 1963.

Forbes, R. J. and Dijksterhuis, E. J., *A History of Science and Technology. Volume 2, The Eighteenth and Nineteenth Centuries*, Penguin Books, Baltimore, Maryland, 1963.

Foreyt, J. P. and Poston II, W. S. C., Diet, genetics and obesity, *Food Technol.*, 51, No. 3, 70, 1997.

Freeman, M., Sung, in *Food in Chinese Culture: Anthropological and Historical Perspectives*, 141, Chang, K. C., Ed., Yale University Press, New Haven, 1977.

Fuchs, C. S., Giovannucci, E. L., Colditz, G. A., Hunter, D. J., Stampfer, M. J., Rosner, B., Speizer, F. E., and Willett, W. C., Dietary fiber and the risk of colorectal cancer and adenoma in women, *N. Engl. J. Med.*, 340, 169, 1999.

Fuller, G. W., *New Food Product Development: From Concept to Marketplace*, CRC Press, Boca Raton, Florida, 1994.

Fuller, G. W., *Getting the Most Out of Your Consultant: A Guide to Selection Through Implementation*, CRC Press, Boca Raton, 1999.

Garcia, D. J., Omega-3 long-chain PUFA nutraceuticals, *Food Technol.*, 52, No. 6, 44, 1998.

Gasser, C. S. and Fraley, R. T., Transgenic crops, *Scientific American*, 266, No.

Gillmor, D., Groceryland, *Business Mag.*, 15, 121, October, 1998.

Goldblith, S. A., 50 years of progress in food science and technology: from art based on experience to technology based on science, *Food Technol.*, 43, No. 9, 88, 1989.

Goldblith, S. A., The legacy of Columbus, with particular reference to foods, *Food Technol.*, 46, No. 10, 62, 1992.

Gori, G. B. and Luik, J. C., *Passive Smoke: The EPA's Betrayal of Science and Policy*, The Fraser Institute, Vancouver, 1999.

Greenwood, J., 'Frankenfoods' uproar grows, *Financial Post*, D3, Feb. 20, 1999.

Han, J. H., Antimicrobial food packaging, *Food Technol.*, 54, No. 3, 56, 2000.

Hartley, D., *Food in England*, Futura Publ., Macdonald & Co. Ltd., London, 1954.

Hasler, C. M., Functional foods: their role in prevention and health promotion, *Food Technol.*, 52, No. 11, 63, 1998.

Hawthorn, J., Too much on my plate, *Proceedings Inst. Food Sci. and Technol.*, 13, No. 4, 281, 1980.

Heinrich, J., Ordinary herb was source of outbreak, *The Gazette, Montreal*, A1, A10, May 11, 1999.

Hoch, G. J., HMR gets healthy, *Food Processing*, 60, No. 5, 88, 1999.

Hollingsworth, P., Web mania: grocers go on-line, *Food Technol.*, 51, No. 12, 22, 1997.

Holmes, A.W., The role of food scientists and technologists in the food industry of the future, *Food Sci. and Technol. Today*, 10, 130, Sept. 1996.

Holmes, R., The obesity bug, *New Scientist*, 167, 26, 2000.

Hoover, D. G., Metrick, C., Papineau, A. M., Farkas, D., and Knorr, D., Biological effects of high hydrostatic pressure on food microorganisms, *Food Technol.*, 43, No. 3, 99, 1989.

Hoover, D. G., Bifidobacteria: activity and potential benefits, *Food Technol.*, 47, No. 6, 120, 1993.

Hulse, J. H., Urban food security: a crisis for creative food systems analysts, *Food Sci. and Technol. Today*, 13, No. 3, 123, 1999.

Idziak, E., Personal communication, 1998.

Innis, M. Q., *Mrs. Simcoe's Diary*, Innis, M. Q., Ed., Macmillan of Canada, Toronto, 1965.

Ishibashi, N. and Shimamura, S., Bifidobacteria: research and development in Japan, *Food Technol.*, 47, No. 6, 126, 1993.

Itokawa, Y., Kamehiro Takaki (1849–1920): a biographical sketch, *J. Nutrition*, 106, No. 5, 583, 1976.

Jay, J. M., Do background microorganisms play a role in the safety of fresh foods?, *Trends Food Sci. Technol.*, 8, 421, 1997.

Jewell, W. J., Resource-recovery wastewater treatment, *American Scientist*, 82, 336, 1994.

Juneja, L. R., Chu, D.-C., Okuba, T., Nagato, Y. and Yokogoshi, H., *Trends Food*

282 ■ Food, Consumers, and the Food Industry

Karel, M., Technology and application of new intermediate foods, in *Intermediate Moisture Foods*, Davies, R., Birch, G. G., and Parker, K. J., Eds., Applied Science, London, 1976.

Katz, F., That's using the old bean, *Food Technol.* 52, No. 6, 42, 1998.

Katz, F., Top product development trends in Europe, *Food Technol.*, 53, No. 1, 38, 1999.

Kaufman, L., Debunking e-commerce, *Financial Post*, D12, May 29, 1999.

Keller, T., New biz, old think, *The Financial Post Mag.*, 69, August, 1999.

Knorr, D., Technology aspects related to microorganisms in functional foods, *Trends Food Sci. Technol.*, 9, No. 8-9, 295, 1998.

Krantz, M., Click till you drop, *Time*, 152, No. 3, 14, July 20, 1998.

Kraushar, P. M., *New Products and Diversification*, Business Books Ltd., London, 1969, Chap. 11.

Kuhn, M. E., Functional foods overdose?, *Food Processing*, 59, No. 5, 21, 1998.

Lachance, P. A., Human obesity, *Food Technol.*, 48, No. 2, 127, 1994.

Lane, S. M., Should universities imitate industry?, *American Scientist*, 84, 520, 1996.

Lee, Y.-K. and Salminen, S., The coming of age of probiotics, *Trends Food Sci. Technol.*, 6, No. 7, 241, 1995.

Leistner, L. and Rödel, W., Inhibition of micro-organisms in food by water activity, in *Inhibition and Inactivation of Vegetative Microbes*, Skinner, F. A. and Hugo, W. B., Eds., Academic Press, London, 1976.

Levins, R., Awerbuch, T., Brinkmann, U., Eckardt, I., Epstein, P., Makhoul, N., Albuquerque de Possas, C., Puccia, C., Spielman, A., and Wilson, M., The emergence of new diseases, *American Scientist*, 82, 52, 1994.

Lewandowski, R., Corporate confidential, *The Financial Post Mag.*, March, 18, 1999.

Lieber, H., Preserved radio-active organic matter and food, U. S. Patent 788 480, 1905.

Lind, M., Regionalism -- who needs it?, *National Post*, A15, March 3, 1999.

Lindsay, D. G., Food chemical safety: the need for a new 'whole food' approach? *Food Sci. and Technol. Today*, 12, 2, 1998.

Lowey, M., Designer crops come under the microscope, *The Gazette, Montreal*, A12, March 8, 1999.

MacKenzie, D., Run, radish, run, *New Scientist*, 164, 36, 1999.

MacKenzie, D., Protein at a price, *New Scientist*, 165, 32, 2000.

Maitland, F. W., *Domesday Book and beyond: three essays in the early history of England*, Cambridge University Press, 1897, Collins Fontana Library, reprinted 1960.

Mallin, M. A., Impact of industrial animal production on rivers and estuaries, *American Scientist*, 88, No. 1, 26, 2000.

Matthews, R., Hidden perils, *New Scientist*, 157, 16, 1998.

Maugh II, T. H., Caution urged in new cystic-fibrosis therapy, *The Gazette, Montreal*, B1, Oct. 11, 1999.

Menzies, D., Checking out the aisles by computer, *The Financial Post Mag.*, 96, December, 1998.

Mermelstein, N., Seeds of change: the Smithsonian Institution's Columbus quincentenary exhibition, *Food Technol.*, 46, No. 10, 86, 1992.

Mertens, B. and Knorr, D., Developments of nonthermal processes for food preservation, *Food Technol.*, 46, 124, 1992.

Messenger, R., The challenge for food processors, *Food Processing*, 59, No. 4, 114, 1998.

Michaelides, J., User perspective, paper delivered in "Expanding the egg horizon" held February 17th, 1997, Toronto, Ontario Egg Marketing Board and the Guelph Food Technology Center.

Moore, J. H., The changing face of university R&D funding, *American Scientist*, 86, 402, 1998.

Morison, S. E., *The European Discovery of America: The Northern Voyages*, Oxford University Press, New York, 1971.

Moroney, M. J., *Facts from Figures*, 2nd ed., Penguin Books Ltd., Harmondsworth, Middlesex, U.K., 1953.

Mote, F. W., Yüan and Ming, in *Food in Chinese Culture: Anthropological and Historical Perspectives*, 193, Chang, K. C., Ed., Yale University Press, New Haven, 1977.

Neff, J., Fat city for fat research, *Food Processing*, 59, No. 2, 31, 1998.

Neusner, N., Obliging the bulge, *Financial Post*, C2, Jan. 4, 1999.

Nguyen, M. and Schwartz, S., Lycopene: chemical and biological properties, *Food Technol.*, 53, No. 2, 38, 1999.

Ohshima, T., Recovery and use of nutraceuticals products from marine resources, *Food Technol.*, 52, No. 6, 50, 1998.

Ondrizeck, R. R., Chan, P. J., Patton, W. C., and King, A., An alternative medicine study of herbal effects on the penetration of zona-free hamster oocytes and the integrity of sperm deoxyribonucleic acid, *Fertility and Sterility*, 71, No. 3, 517, 1999.

Owens, A. M., Herbs linked to fertility risk, *National Post*, A3, March 10, 1999.

Patterson, R. E., Eaton, D. L. and Potter, J. D., The genetic revolution: change and challenge for the dietetics profession, *J. Am. Diet. Assoc.*, 99, No. 11, 1412, 1999.

Pearce, F., Burn me, *New Scientist*, 156, 31, 1997.

Petesch, B. L. and Sumiyoshi, H., Recent advances on the nutritional benefits accompanying the use of garlic as a supplement, *Trends Food Sci. Technol.*, 9, No. 11/12, 415, 1999.

Pszczola, D. E., Take a snack on the wild side, *Food Technol.*, 52, No. 11, 72, 1998.

Ramarathnan, N., and Osawa, T., International conference on food factors: chemistry and cancer prevention. Report on a conference held in Hamamatsu, Japan, December 10–15, 1995, *Trends Food Sci. Technol.*, 7, No. 2, 64, 1996.

Ryu, C. H. and West, A., Development of kimchi recipes suitable for British consumers and determination of vitamin C changes in kimchi during storage, *Food Sci. and Technol. Today*, 14, No. 2, 76, 2000.

Salminen, S., Ouwehand, A., Benno, Y., and Lee, Y. K., Probiotics: how should they be defined? *Trends Food Sci. Technol.*, 10, No. 3, 107, 1999.

Sanders, M. E., Probiotics, *Food Technol.*, 53, No. 11, 67, 1999.

Scherer, C. W., Strategies for communicating risks to the public, *Food Technol.*, 45, October, 110, 1991.

Schmidl, M. K. and Labuza, T. P., Medical foods, *Food Technol.*, 46, 87, 1992.

Scott, A. F., The invention of the balloon and the birth of modern chemistry, *Scientific American*, 250, 126, 1984.

Shortt, C., Report of a meeting "Host-microflora interface in health & disease" May 20–21, 1999, Royal Tropical Institute, Amsterdam, The Netherlands, *Trends Food Sci. Technol.*, 10, 182, 1999.

Silver, H. J., and Castellanos, V. H., Nutritional complications and management of intestinal transplant, *J. Am. Diet. Assoc.*, 100, No. 6, 687, 2000.

Simopoulos, A. P., Diet and gene interactions, *Food Technol.*, 51, No. 3, 66, 1997.

Singer, C., *A Short History of Scientific Ideas to 1900*, Oxford University Press, Oxford, 1954.

Sizer, C. E., and Balasubramaniam, V. M., New intervention processes for minimally processed juices, *Food Technol.*, 53, No. 10, 64, 1999.

Smil, V., China's food, *Scientific American*, 253, No. 6, 116, 1985.

Spears, T., Ottawa discovery: virus destroys cancer tumours, *The Gazette, Montreal*, A1, June 28, 2000.

Speck, M. L., Dobrogosz, W. J., and Casas, I. A., *Lactobacillus reuteri* in food supplementation, *Food Technol.*, 47, No. 7, 90, 1993.

Spitz, P., The public granary, *Ceres*, 12, 16, 1979.

Spring, J. A. and Buss, D. H., Three centuries of alcohol in the British diet, *Nature* 270, 567, 1977.

Stanton, J. L., Marketing or sales? *Food Processing*, 59, No. 7, 51, 1998.

Stelfox, H. T., Chua, G., O'Rourke, K., and Detsky, A. S., Conflict of interest in the debate over calcium-channel antagonists, *New Engl. J. Med.*, 338, 101, 1998.

Stern, C., Live food fight in U.S. pits activists vs. Asian tradition, *The Gazette, Montreal*, F8, December 16, 1998.

Strauss, S., Cyberjournals offer faster, cheaper and fuller research news, *The Globe and Mail*, April 6, 1996.

Streeten, P., Food prices as a reflection of political power, *Ceres*, 16, No. 2, 16, 1983.

Swaminathan, M. S., Environmental and global food security, *Food Technol.*, 46, 89, 1992.

Tannahill, R., *Food in History*, Stein and Day, New York, 1973.

Tedeschi, R., What a tangled Web they weave, *Financial Post*, March 3, 1999.

Thompson, W. I., A. D. 2000, in *At The Edge of History*, Harper & Row, New York,

Toops, D., Americans face a diet dilemma, *Food News*, 34, No. 3, 1, 2000.

Toussaint-Sarnat, M., Alberny, R., Horman, I., and Montavon, R., *2 Million Years of the Food Industry*, Nestlé S. A., Vevey, Switzerland, 1991.

Traill, C. P., *The Canadian Settler's Guide*, first published 1855, reprinted McClelland and Stewart Ltd., Toronto, 1969

Tyler, V. E., *The Honest Herbalist: A Sensible Guide to the Use of Herbs and Related Remedies*, 3rd ed., Pharmaceutical Products Press, New York, 1993.

Vietmeyer, N., Livestock for the landless, *Ceres*, 17, 43, 1984a.

Vietmeyer, N., The lost crops of the Incas, *Ceres*, 17, 43, 1984b.

Walston, J., C.O.D.E.X. spells controversy, *Ceres*, 24, No. 4, 28, 1992.

Wildavsky, A., No risk is the highest risk of all, *American Scientist*, 67, No.1, 32, 1979.

Wilkinson, J. Q., Biotech plants: from lab bench to supermarket shelf, *Food Technol.*, 51, No. 12, 37, 1997.

Williamson, D. M., Gravani, R. B., and Lawless, H. T., Correlating food safety knowledge with home food-preparation practices, *Food Technol.*, 46, 94, 1992.

Winter, C. K., and Francis, F. J., Assessing, managing, and communicating chemical food risks, *Food Technol.*, 51, 85, 1997.

Woodforde, J., *A Country Parson: James Woodforde's Diary 1759 – 1802*, Oxford University Press, New York, 1985.

Worsfold, D. and Griffith, C., Food safety behaviour in the home, *Brit. Food J.*, 99, No. 3, 97, 1997.

Wrick, K. L., Friedman, L. J., Brewda, J. K. and Carrol, J. J., Consumer viewpoints on "Designer Foods", *Food Technol.*, 47, No. 3, 94, 1993.

Young, J., A perspective on functional foods, *Food Sci. and Technol. Today*, 10, No. 1, 18, 1996.

Zee, J. A., Simard, R. E., and Desmarais, M., Biogenic amines in Canadian, American and European beers, *Can. Inst. Food Sci. and Technol.* 14, No. 2, 119, 1981.

Zind, T., Phytochemicals: the new vitamins?, *Food Processing*, 59, No. 11, 29, 1998.

Zind, T., The functional foods frontier, *Food Processing*, 60, No. 4, 45, 1999.

Index

G

H